LITTLE SHIP METEOROLOGY

By the same Author

LITTLE SHIP NAVIGATION (COASTAL)
LITTLE SHIP ASTRO-NAVIGATION
LITTLE SHIP HANDLING (MOTOR VESSELS)

LITTLE SHIP METEOROLOGY

By

M. J. RANTZEN
Lieut.-Comdr. (Sp) R.N.V.R.
Hon. Navigation Instructor, Little Ship Club

BARRIE & JENKINS : LONDON
COMMUNICA-EURUPA N.V.

First published 1961 *by*
Herbert Jenkins Ltd.
3 *Duke of York Street,*
*London, S.W.*1
Second Edition 1974 *published by*
Barrie & Jenkins Ltd.,
24 *Highbury Crescent,*
*London N*5 1*RX*

ISBN 0 214 20110 4

Printed in Great Britain by
Whitstable Litho Ltd., Whitstable, Kent

INTRODUCTORY

This book, the third of my "Little Ship" books addressed primarily to yachtsmen and others who go to sea in ships whose size is measured in tons, rather than in hundreds or thousands of tons, is an attempt to provide the masters of such ships with practical weather wisdom and to enable them to use the Meteorological Office shipping forecasts to best advantage.

That there is real need for such a book I have no doubt. Though always extremely uncomfortable, bad weather may not, perhaps, be a deadly serious matter to the little ship in deep water and with plenty of sea room. It is probable, indeed, that a good ten-tonner, well found and well handled by a good crew, can snug down and ride out in safety anything but a more or less freak storm of quite exceptional severity. It is quite another matter, however, to be caught out in shallow water or with inadequate sea room to leeward, and not a year passes without its toll of lives and ships unnecessarily lost through untimely leaving the shelter of harbour when better weather sense would have led their masters to double up the warps and stay alongside.

Official shipping forecasts, though by no means infallible, are really very good on the average, with a percentage of correctness for short-term forecasts (up to 12 hours ahead) well up in the nineties. Nevertheless, practical use of them in a little ship is by no means easy and I am quite convinced that the common habit of merely listening to that part of a shipping forecast which deals with the area in which you are about to sail is so inadequate as to be dangerous. Taking the advice of some well-meaning blue-jerseyed local of sea-dog appearance is even worse, even if the local in question is really a seaman and not, as so often happens, a bum-boatman in search of a free beer from the innocent yachtsman. What the little-ship master needs is a general picture of the weather and of the way it is changing so that he can properly interpret, for his own area, a shipping forecast which is necessarily general in nature (since it covers a large area in a few short minutes) and can correlate his own observations with what the forecaster has foretold, so as to have a good idea of what is likely to happen if the forecast is a little in error, as it sometimes is. Nine times out of ten such errors are mistakes as to the precise speeds and directions of movement of weather systems and a quite small error of this sort may make a forecast utterly useless to the man who has contented himself merely with listening to the bit about the particular small area in which he happens to be. The same shipping forecast, in the hands of the man who uses it to get a general overall picture and has sufficient knowledge to supplement it by his own observations, can be almost as useful as if it had been completely free of error. For this reason, there is given, in Chapter XI, a method of drawing the weather chart at sea from the forecasts. Though this method appears here for the first time in any textbook, I have practised it myself over a number of years and have found it thoroughly helpful and useful.

The general aim of the book is to impart enough knowledge to enable the little-ship master intelligently to observe the sky, intelligently to anticipate the way in which local conditions are likely to modify a general weather forecast, and to obtain and interpret general weather charts from the official shipping forecasts. This being the aim, there have been very substantial changes in presentation as compared to that of an ordinary, general-purpose textbook on meteorology and, in some respects, very considerable simplification. It has been accepted that the only meteorological instrument likely to be found in a little ship is the barometer—indeed I very much doubt whether any other meteorological instrument is really worth carrying in a ten-tonner—and the only sources of available meteorological information are a broadcast receiver and the master's eyes. Accordingly, observations, such as dew-point and even temperature measurements, which no meteorologist ashore would dream of neglecting, have been, if not exactly discarded, relegated very much to the background, for they are observations which it is usually impossible to take with useful accuracy from an open cockpit only a few feet above the sea. There has been corresponding simplification in the division of cloud and weather into the types to be observed, though I believe that none of this simplification, shocking though it may be to the professional meteorologist accustomed to recognising about sixty different forms of cloud, is at the expense of practical utility in a little ship. In the matter of physics I have kept in mind that my readers will, in the main, be seamen and not scientists and have therefore confined the physics to elementary principles which ordinarily intelligent but not specially trained men will readily follow and which will help them to an understanding of how "the wheels go round" in the mechanism of weather.

In conclusion, I would add that, though I was for some years a naval meteorological officer in an aircraft carrier at sea, drawing synoptic charts every few hours and taking upper air observations and all the numerous instrumental observations necessary to the proper performance of the duties of my office, I have, as a yachtsman over the last quarter century, regularly practised the far simpler methods set out in this book and they have not let me down yet.

M.J.R.

CONTENTS

LIST OF PHOTOGRAPHS

LITTLE SHIP METEOROLOGY

CHAPTER I

WEATHER CHARTS: BROAD PRINCIPLES OF OFFICIAL FORECASTING

Practically speaking, all useful weather forecasting in latitudes in which the weather is variable rather than merely seasonal (as it is in much of the tropics and near-tropics) is based upon the preparation and interpretation of weather charts, or synoptic charts, to give them their more correct name. These are charts on which are printed large numbers of weather observing stations in their correct positions and which are completed every few hours with standardised weather observations taken simultaneously at all the stations distributed over the area the chart covers. Although a certain amount of forecasting can be done on a single observer basis by one man watching the sky around him and noting the behaviour of his own barometer, such forecasting is not only extremely limited but is also very unreliable in any but the most settled of conditions. In variable weather conditions, such as are usual outside the tropics, single observer forecasting cannot, at best, do much more than give a more or less general idea of what is likely to happen in the next hour or two, for the amount of sky within the range of one pair of eyes is very small and soon crossed by a weather system which is moving only moderately fast. In fact, despite the prevalence of longshoremen and self-confident local weather prophets who will cheerfully give a 24-hour forecast on the strength of the colour of the sunset, or the look of the moon or even on the behaviour of cattle, such prophecies are generally so nearly valueless that the practical seaman will ignore them altogether.

Observations for Synoptic Charts. Most of the countries of the world maintain large numbers of fixed meteorological observing stations distributed over their territories. Internationally standardised weather observations are taken at internationally standardised fixed times at all these stations, of which there are about 150 in the United Kingdom and many thousands in

Europe, America, Australasia and the other well-developed countries of the world. At a very large proportion of these stations the times of observations are at 3-hourly intervals. In Europe, for example, the intervals run from midnight to midnight G.M.T.—i.e. at midnight, 0300 G.M.T., 0600 G.M.T. . . . and so on all round the clock. A good many of the stations also take observations at hourly intervals, while a few of the more remote ones only do so at longer intervals—e.g. at 6-hourly or 12-hourly intervals. In addition, standardised weather observations are taken by ships at sea and by aircraft, and, at some special stations, upper air observations are taken by meteorological instruments carried in balloons which also carry radio transmitters arranged automatically to radio the instrumental readings back to earth as the balloons rise. The observations taken at some stations are naturally more complex and detailed than those taken at others, but very large numbers of stations observe the following weather factors:—

Direction and speed of wind,
Barometric pressure,
Amount and nature of change of barometric pressure in the last 3 hours,
Visibility,
Air temperature in the shade,
Reading of a wet-bulb thermometer in the shade,
Proportion of sky covered by cloud,
Amounts and types of low, medium and high clouds,
Weather proper (i.e. whether raining, snowing, and so on), and
Past weather proper in the last 3 hours.

These observations, which are extremely detailed—for example, there are no less than nine possible different reports for type of low cloud alone—are coded into an internationally recognised meteorological code of numbers for quick teleprinter or other telegraphic transmission and the coded messages are sent to distribution centres which redistribute all the messages to all the stations in the network. In this way, in the United Kingdom, for example, each of about 150 stations will receive the weather observations of all those stations within about half an hour or so of their having been taken. Wherever there is a forecaster (as there is at many of the airfields, for example) the coded messages are decoded as they come in and plotted in conventional symbols (of which more anon) on charts on which the fixed stations are already printed in their correct geographical positions.

Drawing of Synoptic Charts. The official forecaster's work really starts from the plotted chart on which the various observations taken at all the stations have been marked, and his first task is to draw in isobars—that is to say, lines joining places where the barometric pressure is the same—in such manner as to "fit" the pressure, wind, and other factors of weather marked in on the chart before him; for, as will be explained in later chapters, many of the different weather factors are inter-dependent and inter-related. He will have before him the last chart he drew and, very likely also, a so-called "pre-baratic" chart, which is really a sort of "prophecy chart" prepared when the last chart was made and showing what the present chart should be like if the weather systems in the last chart have moved and developed in the interval exactly as he calculated they would. Alas for human frailty, a "pre-baratic" is seldom exactly like the actual chart for the time for which it was prepared, but, if his forecast was good (as is usually the case, despite the unkind things said about forecasters whose errors are the only things remembered about them), it will not be seriously different. Indeed, continuity from chart to chart, so that the development from one chart to the next is reasonable and logical, is one of the major things which the forecaster watches for and seeks.

Synoptic Charts. Isobaric Patterns. When the forecaster has drawn his chart he will have before him a pattern or system of isobars, perhaps simple, perhaps more or less complex, which will be of one or other of a number of well-recognised forms, such as that of a depression, which is a system with a low-pressure centre (a "low"), or that of an anti-cyclone, which is a system with a high-pressure centre (a "high").

Fig. 1 is a highly simplified version of an isobaric chart. In this example there is a "high" centred on Ireland and a "low" North of Iceland. The isobars are the lines marked with figures. They may be regarded as pressure "contours"; all points on the line marked 1000, for example, have the same pressure (of 1000 millibars), and similarly all points on any other of the number

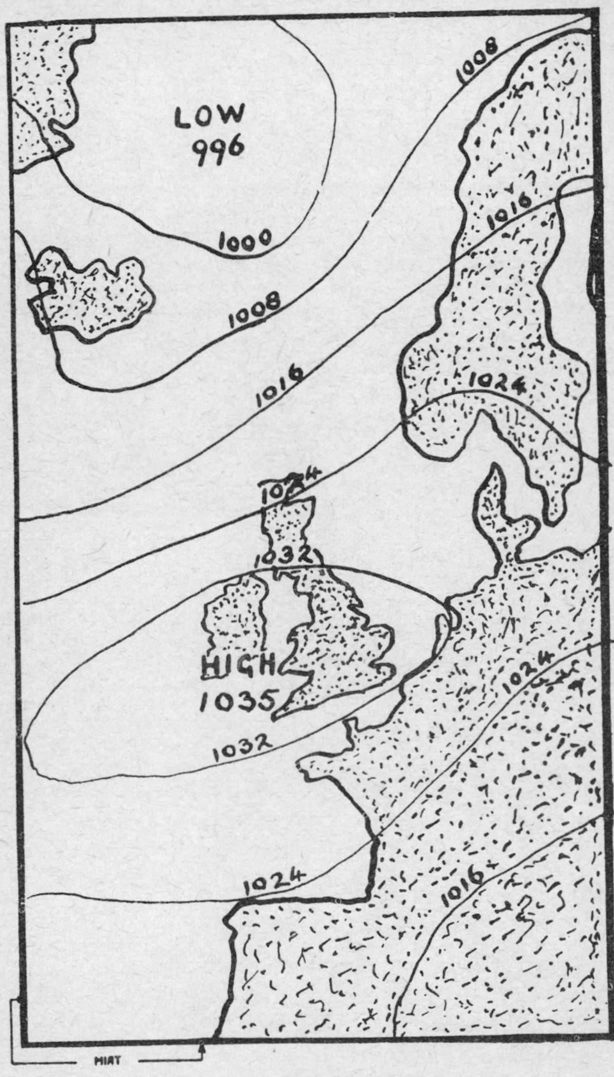
1008
LOW
996
1016
1000
1008
1024
1016
1024
1032
HIGH
1035
1024
1032
1024
1016

FIG. 1.

referenced lines have the same pressure—that marked on the line in question.

Since, in general, different sorts of weather are associated with different forms of isobaric patterns or systems, the drawing in of the isobars takes the forecaster a considerable way towards making his forecast. Careful study of the isobaric pattern, however, and of the weather occurring at different places in the chart, will show considerable areas across each of which is blowing an air stream which, judging from the weather occurring, is all more or less of one type and comes in the same general direction from the same source—for example, the polar regions. Such a more or less uniform air stream is known as an "air mass". Knowledge of the source, origin and past history of an air mass enables the forecaster to prophesy the weather to be expected in those areas over which it will blow during the period covered by the forecast. This is because in those areas the weather will depend on the qualities and properties of the air in the air mass and on the way—usually pretty gradual—in which those qualities and properties become modified as the mass moves on over land and sea. Thus, to give a very simple example, an originally very cold air mass, blowing with a North wind from polar regions southwards over the British Isles in summer, will gradually warm up as it moves southward over the heated land, and therefore its properties, and the weather to be expected from it, will change correspondingly and gradually. Recognition of air masses, their general properties, and their behaviour from the weather point of view, constitute a major factor in successful forecasting.

Synoptic Charts. In a great many cases, however, and in practically all cases in which the isobaric pattern is that of a depression, the recorded observations on a chart will show weather changes, as between some stations and other stations quite near them, which are altogether too abrupt to be satisfactorily accounted for by a consideration of the properties of a single air mass alone. These changes will appear to occur across more or less well-defined lines or belts extending across the chart or partly across it, and the weather at stations on and near such a line will be reported as quite different from that at stations well to one side of the line which, in turn, will report weather quite different from that reported by stations well to the other side of the line. Fig. 2 illustrates this sort of thing. On this particular weather

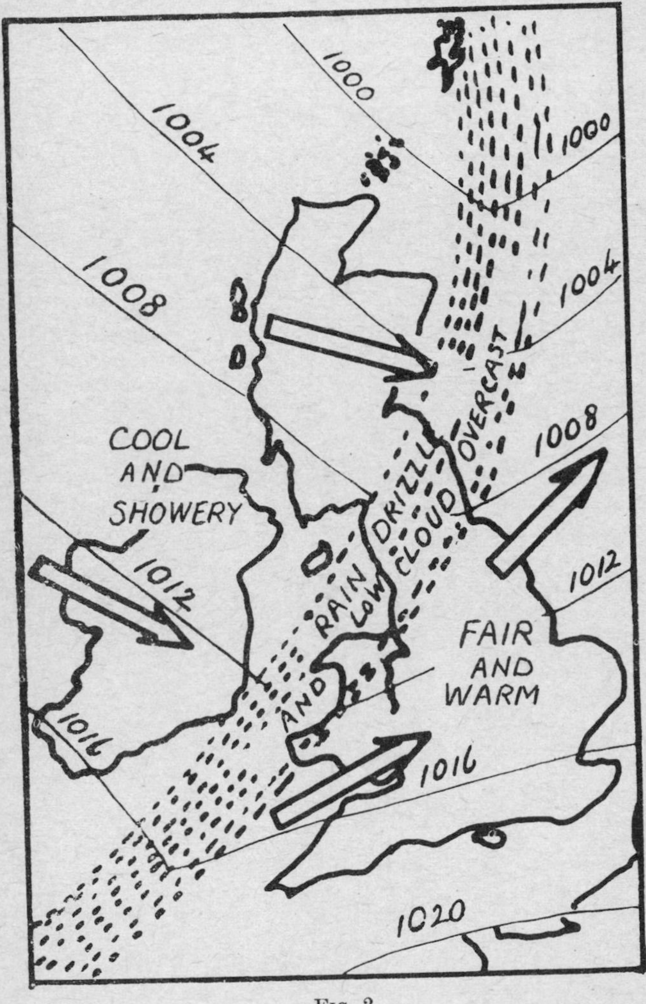
1000
1004
1000
1008
1004
OVERCAST
1008
COOL
AND
SHOWERY
DRIZZLE
CLOUD
LOW
RAIN
AND
1012
1012
FAIR
AND
WARM
1016
1016
1020

FIG. 2.

map there is a well-defined rain belt (indicated by the dotted area) stretching away in a generally South to South-West direction, from the neighbourhood of Shetland, across Northern England into the Western approaches. Stations in this belt are reporting rain or drizzle and much very low cloud, generally overcast. Stations East and South-East of this belt are reporting South-West winds with warm temperatures, fair weather and broken cloud except quite close to the belt. Stations to the West and North-West of the belt are reporting West to North-West winds and markedly lower temperatures with showers, heaviest near the belt, and large but separated clouds. Moreover, when the forecaster has been drawing his isobars he has found that, in order to make them "fit" the pressures recorded by the various stations, he has had to make them change direction more or less sharply where they cross the belt.

The comparatively sharp differences between the weather reports from stations to the East of the belt and those to the West of it are an indication that the former stations are being blown over by one air mass and the latter by another and different one: in other words, that one air mass is being replaced by another along the belt. In fact this is what is happening, and lines along which one air mass is being replaced by another are known as "fronts". The forecaster marks such fronts on his chart, for their detection and recognition are an extremely important part of forecasting. Where, as in the example just mentioned, a cold air mass replaces a warmer one, the line of demarcation is called a cold front. Similarly, where a warm air mass substitutes itself for a colder one, the line of demarcation is called a warm front. A perhaps unexpected property of air masses in general is the extent to which each of two air streams will retain its own character and properties without much mixing even when moving in juxtaposition to one another. It is rather like running water simultaneously from the hot and cold taps into a large bath: some of the water in the bath will be quite hot and some quite cold and a good deal of vigorous and maintained stirring is required to bring all the water, even temporarily and approximately, to the same temperature. Because of this power of air masses to preserve their own identities, when a warm air mass meets a colder one, the warmer air, being lighter, will climb up over the colder if it is coming up behind it (as occurs at a warm front) without much mixing. Similarly, if a cold air mass comes up behind a warmer one (as it does at a cold front) it will, being heavier, cut in under the warmer air, again without much mixing. In both cases, the warmer air is lifted almost as if there were an impenetrable boundary between the two air streams. This action of one air mass riding up over or cutting in under another is characteristic of fronts and of very great importance in forecasting, because if air is lifted through a great height it suffers great pressure changes which cause it to produce weather phenomena quite different from those that would have been produced had no lifting taken place. On the opposite sides of a front, therefore, we will have what may be termed different "air mass weathers"—i.e. different weathers due to the different properties of the two air masses which meet at the front. At the front itself, however (that is to say, over the boundary surface where the air streams adjoin), the weather will be what may be termed "frontal" or "change of air mass" weather—i.e. weather due mainly to the fact that the meeting of the air streams has caused the warmer and therefore lighter air to be lifted in riding up over the heavier and colder air. That is why, in Fig. 2, the weather in the belt itself is quite different from the weathers due to the air masses on the two sides of it.

Local Weather Variations. The considerations so far set forth have been of the most general and preliminary kind—merely broad indications of the way in which forecasting is dependent upon the recognition and identification of different pressure systems ("lows", "highs" and so on); recognition and identification of different air masses; and recognition and identification of fronts. More specific and detailed information on these matters and the ways in which the movements of pressure systems and fronts can be foretold will be given in later chapters. It may be remarked here, however, that when all this has been done there will still remain local weather modifications, which are often of great importance to little ships, especially when near land, due to local topography and things of that kind. Once the main forecasting has been done, the foreseeing of local modifications is largely a matter of common sense; for example, if a S.W. gale is foretold for the English Channel it does not require a meteorological genius to foretell that there will not be a gale, but, on the contrary, good shelter, close to the East of, and under the lee of, a well-defined headland like Dungeness. Some of these local modifications are not, perhaps, obvious at first sight and will be dealt with in later chapters.

The Shipping Forecasts and the Little Ship. From the foregoing general description of the mass of information available to the official forecaster and of the way he uses it, it will be at once apparent that the master of the little ship, with no more meteorological equipment than his barometer, his broadcast receiver for receiving the shipping forecast, and his eyes, has no chance at all of setting himself up in competition with the official forecasters, and he is a fool if he tries. Certainly this book does not seek to do anything so silly as to encourage him to make his own forecasts independently of the official meteorological services. If, however, he has a reasonably good grasp of the foundations of meteorology, understands in principle how to draw and interpret weather charts, and knows what to look for in the sky and in the readings of his own barometer, he will be able, quite quickly and easily, not merely to listen to the shipping forecasts but to use them at sea to draw his own weather charts which, though naturally nothing like as detailed or as good as those the official forecaster has ashore, will, nevertheless, be practical and useful charts and good enough to be of great value to him, enabling him to obtain vastly enhanced benefit from the official forecasts and not to be taken by surprise when, as sometimes happens, they go a little wrong in their timing. The aim of the chapters which follow is to impart this knowledge and understanding so as to enable such enhanced benefit to be obtained.

CHAPTER II

OBSERVING AND LOGGING: BAROMETER, WIND AND SEA

Since, in the chapters which follow, there will be many illustrations in which conventionalised symbols are used to represent weather observations, it is convenient, at this stage, to give recommended observations to be taken at sea and recommended symbols for recording them in the ship's log without the wastage of time and space which would accompany attempts to record them in words. The professional meteorologist has an elaborate, internationally recognised, official code of symbols which he uses to record the extremely detailed observations he makes. For obvious practical reasons, it is quite impossible for the master of the little ship to make observations in anything like the detail used by the professional weather man, even if he had the knowledge and training to do so, and in consequence the official code is far too complex to be of practical use to him. There is, however, great advantage to be obtained from routine observations made regularly at sea and with a good deal more detail than is commonly the case in yachts, and an easily memorised, simple code of symbols for recording such observations quickly and in a small space in the ship's log is most useful. Such a code will be given in this and in the next chapters. The professional meteorologist will at once recognise it as a highly simplified version of the official code, but it is sufficient for the practical purposes for which it is intended.

Temperature Measurements. In aircraft carriers and major warships generally it is customary to measure air temperatures and wet-bulb temperatures in the shade, and also sea temperatures. Unfortunately, owing to practical circumstances in the average yacht, temperature measurements for weather purposes must be regarded as impracticable and, with few exceptions, not worth the trouble of taking. Although one often encounters air-temperature measuring thermometers in yachts, they are really little more than ornaments. Simple though a thermometer is, the difficulty of finding a place to mount it in a really small vessel where its readings will not be merely misleading is almost insuperable. A thermometer in the cabin is obviously quite useless to indicate the temperature in free air, and if it is put in the cockpit, or on deck, or in the wheelhouse—even if it is mounted in a proper screen—it is so likely to be affected by spray or moisture, or by strictly localised pockets or streams of air which have been cooled by spray or heated by some warmed part of the ship, that its readings will be quite unreliable. Similar remarks apply, with even greater force, to wet-bulb thermometers and hygrometers; and while sea temperatures can be observed if one is prepared to go to the trouble of taking samples of the sea from undisturbed water a few feet down, the value of sea temperatures to a man who cannot get reliable air temperatures or wet-bulb temperatures is so nearly zero (except possibly for the cruder navigational purposes, such as finding out if one is still in the Gulf Stream) that they are really not worth bothering with.

Aneroid Barometer. Ninety-nine times out of a hundred the only practically useful meteorological instrument for the little ship is the aneroid barometer. Fig. 3 illustrates the principle of this instrument. It consists essentially of a partially evacuated,

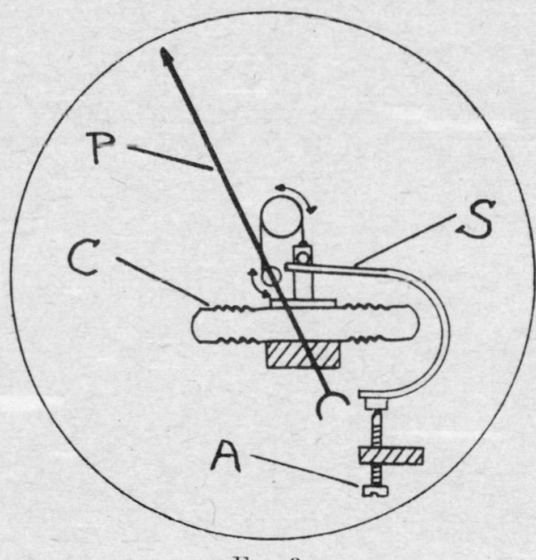

FIG. 3.

hermetically sealed capsule C with flexible walls, and which is exposed to atmospheric pressure which accordingly tries to press the walls together. This is resisted by a powerful spring S, shown as a bar-spring of more or less U-shape, which balances the atmospheric pressure and prevents the capsule walls being collapsed together. Accordingly, the distance between the walls varies with the atmospheric pressure. One of the walls is anchored while the other is mechanically connected in some suitable way to a pointer P which moves over a scale calibrated in units of pressure. A screw A enables the position of the needle on the scale to be adjusted. In the usual forms of instrument the head of the screw A is accessible through a hole in the back or side of the casing of the barometer. There is usually a subsidiary pointer (not shown) for "holding" a previous reading.

Millibars. Inches. Millimetres. The modern unit of atmospheric pressure, now used all over the world, is the millibar, and if acquiring a new barometer it is best to get one with a scale marked in millibars. There are still, however, many aneroids having scales marked in inches of mercury (the height in inches of a column of mercury which the atmospheric pressure will support) or, in countries using the metric system, in millimetres of mercury. Conversion from inches or millimetres to millibars (mbs.) is simply a matter of arithmetic, since 1016 mbs. equals (near enough) 30 inches or 762 millimetres of mercury. Fig. 4 is a convenient conversion scale, for the use of those having aneroids with inch scales, which will convert inches to millibars by direct inspection.

Setting the Barometer. A minor defect of the aneroid barometer is that the spring tends slowly to "settle down", and in the course of a season a barometer which started by giving correct readings is apt slowly to become more and more in error. This is not so serious as it sounds, since, as will be seen later, changes in the barometer are more important than the actual readings themselves. Nevertheless it is wise to keep the barometer reading correctly in order that its indications may be correlated properly with the barometric readings for various stations included in the shipping forecasts. It is normally sufficient to adjust the barometer by means of the screw A once, or possibly twice, during a season. The thing to do is to ring up a nearby major weather station, such as a main airfield (the nearer the better), and ask for

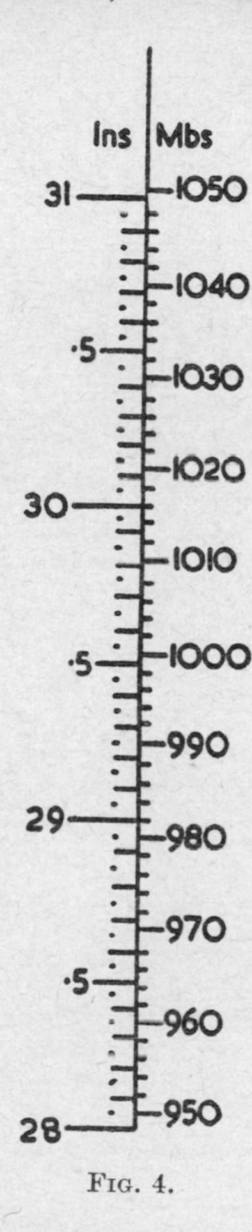

Fig. 4.

the weather officer on duty. Tell him your position and that you want to adjust your barometer and ask him for the "M.S.L." (mean sea level) pressure in millibars or inches or whatever your barometer scale is marked in. Most major airfields have a meteorological officer and, in my experience, they are invariably helpful and most willing to give you the information you want. If you do not know of a nearby airfield and cannot find one in the telephone directory, you can get the information by phoning the nearest weather bureau—for example, in the United Kingdom, the meteorological service in London. Having got the information, adjust your instrument by gently turning the screw A to give the same reading. Try to turn the screw without pressing on it, otherwise you will probably find the reading "jump" when you take the screwdriver away. This adjustment will probably last the whole season. *It is important to make the adjustment when the instrument is actually in the ship: do not take it ashore and try to do it there; for if you are above the level normally occupied by the instrument in the ship it will read high when you get it back aboard, to the tune of nearly 1 mb. for every 29 ft.* (*approximately*) *you have lowered it since you set it.*

Reading and Logging the Barometer. It is recommended to read the barometer every 3 hours and to set the auxiliary hand anew at each reading. Since the full range of atmospheric pressure at sea level is (practically) from 950 mbs. to 1050 mbs. it is sufficient to record only the tens and units figures, and the first decimal figure. Thus a reading of 1012·8 can be recorded as 12·8: it cannot have been 912·8. Similarly 88·2 can mean only 988·2. Also record the amount of change since the last reading, together with the nature or type of change which has been occurring. It is wise, for practical purposes, to differentiate between six types of change, namely:

1. Falling logged thus ⟍ (wavy)
2. Falling then steady ,, ,, ⟍_
3. Falling then rising ,, ,, \/
4. Rising ,, ,, ⟋ (wavy)
5. Rising then steady ,, ,, /‾
6. Rising then falling ,, ,, /\

Reading and Logging Wind. Beaufort Scale. Wind should be observed and logged in direction and force, the former in compass points, true, and the latter in accordance with the Beaufort Scale which is detailed in the table below. Wind direction is best observed by looking at the sea in open water and noting the run of the wave fronts and, when they can be seen, the direction of the "wind lines". The wave fronts (waves, *not* swell) in open water extend nearly always at right angles to the wind direction, and the "wind lines" can frequently be plainly observed as lines or streaks extending along the slopes of waves in the direction of

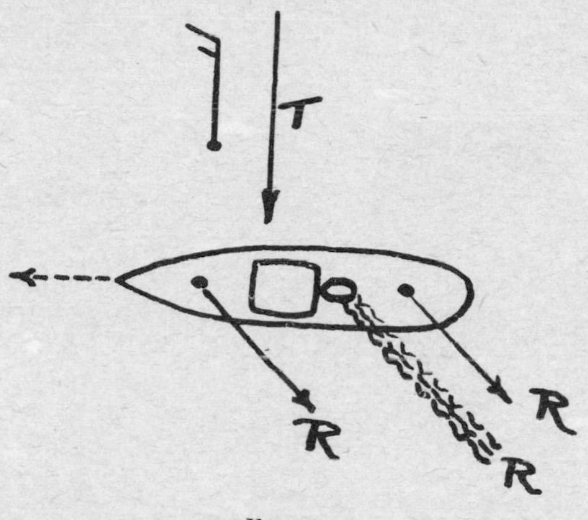

FIG. 5.

the wind. Often the observation is best taken looking down-wind, rather than up-wind. *Do not look to your burgee or exhaust smoke for wind direction*, for this only gives wind *relative to the ship;* and if the ship is moving fast or the wind is light, the relative wind direction and the true wind direction can be very different.

This effect is clearly shown by the example in Fig. 5 which depicts a ship steaming W. at 10 knots with a 10-knot North wind T dead on her beam. Due to the combined effect of her own motion and the wind, her burgee and flags, and her funnel smoke, will not stream out S. with the true wind, but, as shown at R, in a S.E. direction, thus giving the impression of a N.W. wind.

Judgment of wind strength is also best effected in a little ship by looking at the sea, again usually looking down-wind. Though a number of quite good wind-speed measuring instruments which can be held in the hand are now available, it is at least very doubtful whether they can be used reliably in really small craft because of the nearness to sea level, back draughts from sails and that sort of thing. The following table gives the different "forces" of the Beaufort Scale, together with the corresponding usual descriptions; the wind speeds in knots; the accompanying states of sea in open, deep water; the approximate heights of waves to be expected, also in open deep water; the probable effects on the average 10-ton yacht; and the symbols used to represent the scale numbers. Accurate judgment of wind speeds in a small vessel is very difficult—there is a marked tendency to over-estimate head winds and under-estimate following winds—but the "state of sea" column will be found to give a good practical guide.

Nowadays the Beaufort Scale goes up to Force 17, wind speeds between 64 knots and 118 knots being divided in steps between Forces 12 and 17. It is considered that no useful purpose is served by detailing Forces 13 to 17 and that the master of the little ship will be adequately satisfied—if satisfied is the proper word—with knowing that a wind, forecasted or existing, is over 64 knots!

As regards the symbols, it will be seen that they are all "feathered" arrows, with a whole feather for each two steps in the scale and a "half" feather for each intervening one step. The arrows "fly with the wind". Thus the wind symbol in Fig. 5 denotes a wind of Force 3 blowing from the North (true-wind directions are always given in "true"—i.e. with reference to geographical North). If the wind is unusually gusty, record it by an arrowhead on the side of the "shaft" of the arrow, thus: . Because of the effect of the sea in intermittently shielding a small ship by the waves themselves, winds of about Force 6 or 7, or more, are almost invariably gusty as experienced by a ship of around 10 tons.

It should be noted that the states of sea and approximate wave heights in the table are for open sea, in deep water, where there is a long stretch of sea to windward—"fetch" as it is called—and

Beaufort Scale No.	Description	Wind Speed in Knots	State of Sea	Probable Open Sea Wave Height (feet)	Probable Action in 10-tonner	Symbol
0	Calm	0	Flat and glassy	Zero	No steerage way	None
1	Light air	1–3	Ripples	–	Just "ghosting"	
2	Light breeze	4–6	Wavelets	up to 1	Light canvas	
3	Gentle breeze	7–10	Large wavelets. Occasional white horses	2–3	3–4 k full and bye	
4	Moderate breeze	11–16	Frequent white horses	3–5	Good sailing breeze. Well heeled on the wind.	
5	Fresh breeze	17–21	Plentiful white horses. A little spray.	6–9	One reef in main.	
6	Strong breeze	22–27	White everywhere. Fair amount of spray.	9–13	Two reefs in main.	
7	Moderate gale	28–33	Heaping sea. Some white streaks along wind direction.	13–18	Reefed right down or storm canvas.	
8	Gale	34–40	Crests blow off waves. Well marked plentiful white streaks in wind direction. Some spindrift.	18–24	Hove to	
9	Strong gale	41–47	Dense white streaks. Spindrift fills lower air, impairing visibility.	24–30	Hove to	
10	Whole gale	48–55	Sea indescribable when viewed from a small vessel. If unlucky enough to be at sea she will be battened down and under bare poles.			
11	Storm	56–63				
12	Hurricane	63 and up.				

where there is either no tidal stream or a stream whose speed is small compared to the wind speed. In shallow water, the sea can be expected to be worse and more dangerous than in deep water —the waves being usually shorter, steeper and breaking earlier. If there is little or no "fetch", as is the case with an off-shore wind, the sea will be much reduced. A strong tide against the wind will increase and steepen the sea, and one with the wind will have the opposite effect. Thus a tidal stream running at 3 or 4 knots as against a Force 5 wind will produce a state of sea normally appropriate to Force 6. Of course, heavy rain will produce a certain amount of "flattening" of the sea, but this effect is not very great and manifests itself mainly in a small reduction of the amount of white about, without greatly affecting the height of the waves.

1. Fair weather Cumulus. (Ijsselmeer: summer)

2. Large Cumulus (Mediterranean: autumn)

3. Cumulo-Nimbus with Anvil (N. Atlantic: winter)

4. Strato-Cumulus (The Sound: summer)

5. Thick Stratus and Scud. Force 9. (Western approaches: winter)

CHAPTER III

OBSERVING AND LOGGING: CLOUD, WEATHER AND VISIBILITY

Amount of Cloud Cover. It is often useful to record the amount of sky covered by cloud (of any sort) since this has a considerable bearing on whether land or sea breezes can be expected. For practical little-ship purposes it is sufficient to note one or other of six possible conditions, namely: (1) no cloud at all, (2) sky not more than $\frac{1}{4}$ covered, (3) sky $\frac{1}{4}$ to $\frac{1}{2}$ covered, (4) sky more than $\frac{1}{2}$ covered but not completely so, even if only showing through in a few small places, (5) completely overcast, and (6) sky hidden from observation—as, for example, by fog or snow or in rain on a dark night. Recommended symbols for these six possible conditions are in the form of different numbers of lines in the circle by which the observing station (your ship) is normally represented. Here are the six recommended symbols for logging:

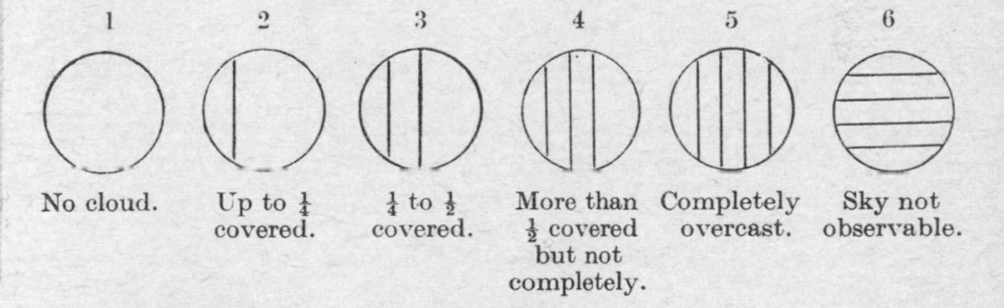

1	2	3	4	5	6
No cloud.	Up to $\frac{1}{4}$ covered.	$\frac{1}{4}$ to $\frac{1}{2}$ covered.	More than $\frac{1}{2}$ covered but not completely.	Completely overcast.	Sky not observable.

Cloud Observations. It is asking too much to expect any but a trained and experienced meteorologist to estimate cloud heights at sea or even to distinguish reliably between low and medium clouds, but it is, in general (though, alas, not always), easy enough to distinguish between very high cloud of the cirrus types and the very much lower medium and low clouds. It is, therefore, recommended not to attempt any height distinction other than that between high (cirrus) cloud and the rest. As regards cloud types, the important thing is to distinguish between clouds which have vertical development and those which are mere shapeless masses, for this distinction is, as will be seen later, extremely important. Clouds with vertical development—the shapely, curved clouds, usually more or less individually separated, which range from the floating "puffs of cotton-wool" that one sees on a fair summer's day to the towering thunder-cloud with an "anvil"—are known as cumulus clouds. The shapeless, formless, flat, spreading clouds, which look more or less like fog up in the air—which is what they generally are, by the way—are stratus clouds. There is an intermediate general type, neither one thing nor the other, which looks like stratus with some shape in it, usually wave-like, called strato-cumulus. It is very common in British and Northern European waters and is sometimes called North Sea cloud, for obvious reasons. It is usually thick and darkish grey in colour when seen from below.

Cloud Symbols. The recommended symbols for clouds are more or less like the clouds they represent, those for cumulus clouds being largely curved. High cloud symbols are put immediately above the "station circle" and symbols for lower clouds are put immediately below it. If, as not infrequently happens, there is more than one type of lower cloud or more than one type of high cloud simultaneously present, it is recommended to log only the dominant type in both cases, or, if this is in doubt, the types which seem to be becoming dominant.

Types of Lower Cloud. It is adequate, for practical purposes at sea, to recognise half a dozen types of lower cloud. Some of these types—and especially the stratus types—are extremely difficult to portray in pictures because of their very shapelessness, but some of the main types with their descriptions, and symbolic representations, are shown in the photographs facing pages 32, 33, 48 and 49. In the example of cumulus of fair weather, note particularly the individually detached nature of the clouds and that they have plainly observable vertical development, though not very much. Compare this with the pictures of large cumulus. This is obviously of the same general nature as cumulus of fair weather, but the vertical development is much greater and there is a pronounced "cauliflower" or "boiling up" look. With this type of cloud the individual detachment may sometimes be as great as that of fair-weather cumulus, though where there are a lot of clouds the detachment may hardly be visible when seen from below. If the vertical development is enough, thunder-cloud (see the plates) is formed. This type of cloud sometimes has enormous vertical development—it may tower up to 20,000 feet or more—and frequently terminates at the top in an "anvil", a term which explains itself. It is associated with thunderstorms, and if rain is falling from it (as is commonly the case) the technical term used for it is cumulo-nimbus. Frequently there is "scud"—i.e. broken pieces of storm-cloud—beneath it. The anvil tops are commonly feathery in appearance, and bits which blow off, as they often do, can easily be mistaken for cirrus. Care should therefore be taken before detached bits of high white feathery cloud, seen in the sky when thunder-cloud is about, are logged as cirrus. They may look like cirrus, but are probably not.

One of the plates shows a good typical specimen of strato-cumulus. It is one of the commonest forms of cloud in British waters. Commonly it is thick, dark grey in appearance and extends over wide areas without a break, or nearly so. It looks much like stratus but differs from it in that there are parts with quite definite shapeliness—usually wavy, as in the photograph.

Stratus is characterised by a complete or almost complete absence of any form or shape, though there is often fragmented storm-cloud—"scud"—beneath it. One of the photographs shows—so far as a photograph can—a sky which is completely overcast with thick stratus. It is usually darkish grey in colour, and, if thick, quite dark. If it is thin, the sun or moon can be detected through it, though the outline will be blurred—"watery sky" is a common appearance produced by thin stratus. Thick stratus completely hides sun or moon.

The recommended symbols for these six forms of lower cloud are as shown below and all are put under the "station circle":

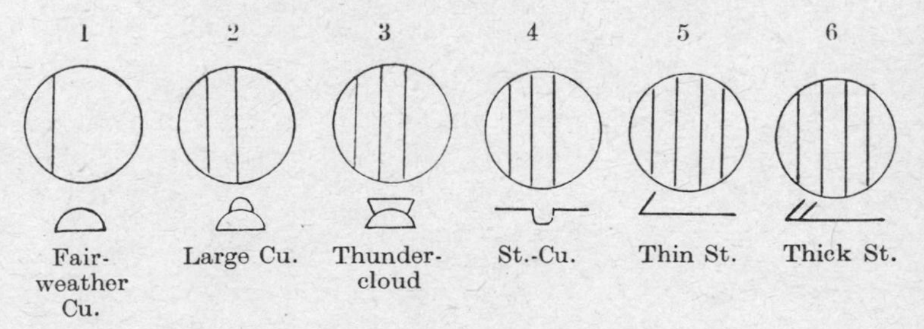

Types of High Cloud. From the forecasting point of view, one of the most important types of high cloud to notice is cirrus. This is the familiar feathery, white or nearly white cloud, obviously at a great height, which has been called "mare's-tails" by generations of seamen. It is one of the first visible signs of an advancing depression; but since there are other meteorological conditions which can cause its appearance, it is as well to note whether it is increasing. Our first two types of high cloud are, therefore, "cirrus" and "cirrus, plentiful and increasing". One of the plates shows a good specimen of "cirrus, plentiful and increasing" from which the feathery "mare's-tails" or "wind-swept" look can be clearly seen.

The second important type of high cloud is cirro-stratus, which follows cirrus during the onset of a depression but which can also form due to other causes, so it is again useful to note whether it is increasing or not. Our next two types of high cloud are, accordingly, "cirro-stratus" and "cirro-stratus, plentiful and increasing". Cirro-stratus is a thin, formless whitish veil which, if increasing, spreads over the whole sky giving the "grey sky" of the seaman. Like cirrus, it is actually composed of ice and the sun or moon can usually be seen through it with an edge clear enough to take a sight, if, of course, there is no lower cloud in the way. Because it is of ice crystals it often gives rise to a halo (a luminous ring) round the sun or moon, the diameter of the halo being a little under 45°. In fact, if a halo is seen round the sun or moon, cirro-stratus is certainly present and should be logged even if it is night and the cloud itself cannot be seen. Cirro-stratus, like stratus, is extremely difficult to show in a picture, but it can be seen in one of the plates stretching up from the sea on the far side of cirrus and, in another photograph, above large cumulus over land.

The last of the types of high cloud which it is as well to distinguish is cirro-cumulus. It is not one of the commoner clouds in British waters and is not of very great importance from the forecasting point of view. It is of broken appearance, whitish, and appears as a rule in the form of large numbers of fairly closely packed white globules, frequently in rows. The well-known "mackerel sky" is an example of cirro-cumulus. It is shown in the middle part of one of the photographs, seen through gaps between large cumulus clouds far below it.

The recommended symbols for these five types of high cloud are all put, as shown below, over the station circle:

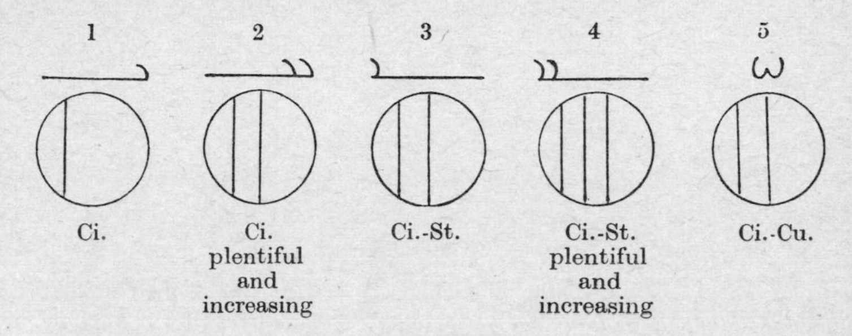

Weather Proper. Observation and logging of weather proper can conveniently and practically be simplified down to a mere eight types, namely: drizzle, rain, snow, showers, thunderstorm, haze, mist (which can be taken as visibility between ½ and 1 sea mile), and fog, which is visibility of ½ a sea mile (1000 yards) or less. Here are the eight recommended symbols, which are pretty obvious and self-explanatory:

1	2	3	4	5	6	7	8
Drizzle	Rain	Snow	Showers	Thunderstorm	Haze	Mist	Fog
,	•	*	▽	R	∞	=	≡

Sleet can be represented as rain and snow, thus: •/*

The first three—drizzle, rain, and snow—can be regarded as being either intermittent and light, intermittent and heavy, continuous and light, or continuous and heavy. If intermittent and light, the symbol is logged as shown above. If intermittent and heavy, the symbol is repeated vertically—e.g. intermittent heavy snow is logged thus: ⁑. If continuous and light the symbol is repeated horizontally, thus • • for continuous light rain. If continuous and heavy, the symbol is repeated both horizontally and vertically, in a triangle, thus ∵ for continuous heavy rain.

Visibility. This is best observed and logged by a figure representing the number of sea miles visibility if of 1 mile or above; by the fraction ½ if between ½ a sea mile and 1 sea mile; or by a figure representing the number of yards visibility if below ½ a sea mile. It is important, for confident ship handling, to know the visibility if it is below about 1 sea mile, but it is extremely difficult to judge it when out of sight of land and when the visibility is low enough for the horizon not to be seen. It is wise, therefore, in misty conditions, to take any opportunity of measuring it which may occur. One useful but often overlooked method of measurement which often occurs is to read the log when passing a light-vessel, buoy or other fixed object, watch it until it is just about to fade out, and then note the difference of the two readings to ascertain the visibility. Another often useful trick is to estimate the speed of a steamship which passes you on your course, or on the reciprocal course—an experienced man can usually make a fair estimate of a steamer's speed at a glance—read your watch when abreast of her, and read it again when she fades out. If you know your own speed, it is simple mental arithmetic to calculate visibility from the relative speed between your own ship and the passing vessel, remembering that 1 knot is (near enough) 100 ft. per minute. Thus, if you are doing 5 knots and are passed by a ship steaming at 8 knots in the same direction, the relative speed is 3 knots and she is leaving you at 300 ft. per minute. If you lose her ahead in 15 minutes, the visibility is 15×300 ft. $= 1500$ yards $= \frac{3}{4}$ mile (nautical).

The Complete Weather Log Entry. This completes the useful practical observations, which are: wind direction and speed; total amount of sky covered; type of high cloud; visibility; weather proper; barometer and type and amount of change of barometer; and type of lower cloud. By putting the appropriate symbols always in the same standardised positions round the station circle it is easy, once you have memorised the code—and it really requires very little memorising—to record a quite full set of observations in a very small space in the log and quicker than it would take to say them, let alone write them, in words. The recommended standardised positions are: wind direction and speed by the angle of the feathered "wind arrow" to the station circle; total amount of cloud cover by lines in the circle; type of high cloud at position 12 o'clock on the circle; visibility and weather proper at 9 o'clock, with the visibility figure on the left of the two; barometer, type and amount of change of barometer at 3 o'clock, with barometer on the left of the three entries, type of change in the middle and amount of change on the right; and type of lower cloud at 6 o'clock.

Here is an example which shows the completeness, simplicity and neatness of the method of logging: Wind, S.W., force 5 unusually gusty; sky half covered; visibility 5 miles; intermittent rain; barometer 997·6 mbs., falling, dropped 3·2 mbs. in past 3 hours; cirro-stratus, plentiful and increasing; lower cloud mainly thin stratus. This considerable mouthful, recorded as above recommended, would appear in the log as the following, almost self-explanatory, "weather entry":

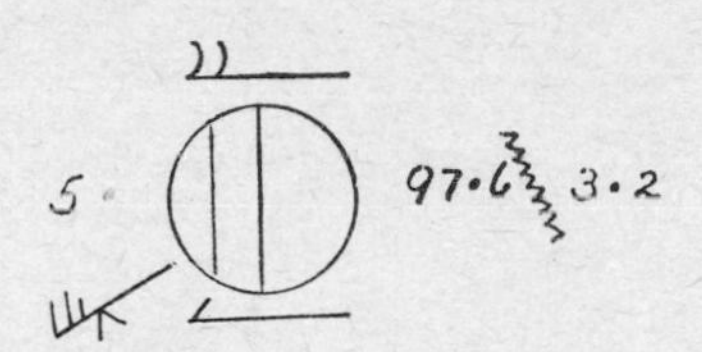

Keeping Track on the Barometer. Every seaman knows the importance of keeping a close eye on what the barometer is doing, and merely inserting figures of pressure in the log is not really a satisfactory way of doing this. It is strongly recommended to keep an ordinary barograph chart, the more open scaled the

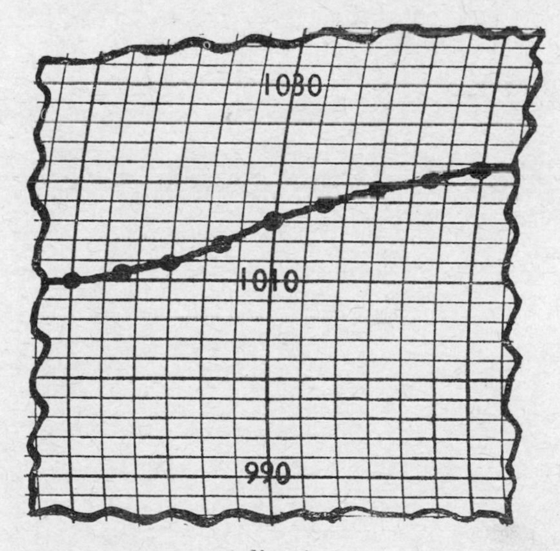

FIG. 6.

better, printed in whatever units of pressure your aneroid scale is marked with, in a transparent closed envelope of Perspex or similar material so that the chart can be seen through. Mark each barometer reading you take on this chart by a dot or cross made with china-glass pencil on the Perspex so that it appears in the correct position against the chart. The dots or crosses can then be joined by a smooth curve in china-glass pencil, though in fact they will be close enough together to give a satisfactory impression of a curve, especially if dots are used, even if they are not joined. The result, which is a useful approximation to the track made by a barograph, is illustrated by Fig. 6, which shows part of an ordinary barograph chart recorded in this way.

The normal barograph chart extends over a week of time so that, when you are approaching the end of a week on the chart, you can wipe out the china-graph marks for the first few days and start again. As will be at once apparent, a record of this sort really does show what the barometer is doing—much more so than mere figures in a log.

CHAPTER IV

ISOBARS AND WIND

Pressure Gradient. An isobar is a line joining places having the same barometric pressure. An isobar on a weather chart is thus quite closely analogous to a height contour on an ordinary map. Movement across a map from a low-level contour to a high one, or vice versa, involves movement up or down a gradient; and the closer together the contours are, the steeper is the gradient. Similarly, movement across a weather chart from a low-pressure isobar to a high-pressure isobar, or vice versa, involves movement along a *pressure gradient*; and the closer together the isobars are, the greater is the gradient. It is obvious that if the earth were stationary the existence of a pressure gradient would tend to cause the air to move down it, just as the existence of an ordinary gradient would tend to cause any body which was free to move to roll down it. One would, therefore, at first sight, expect winds to blow directly from a place of high pressure to a place of low pressure at a speed which would depend on the difference of pressures between the two places, and there is, in fact, a force which does act in this way.

Gradient Wind Force. The gradient wind force is the force above referred to which, if acting alone, would cause winds to blow directly from a "high" to a "low". Clearly the direction in which it acts is at right angles to the isobars.

Thus in Fig. 7 there are gradient wind forces outwardly from the high-pressure centre shown over the British Isles in directions, two of which are indicated by the arrows, which are along the pressure gradients which exist between the high and the lows to North and South of it. Were these forces the only ones acting on the air the winds would blow directly out of the high at speeds depending on the gradient. In fact, however, this does not occur because the gradient wind force is not the only force acting. There is a second force, called the geostrophic force, which is due

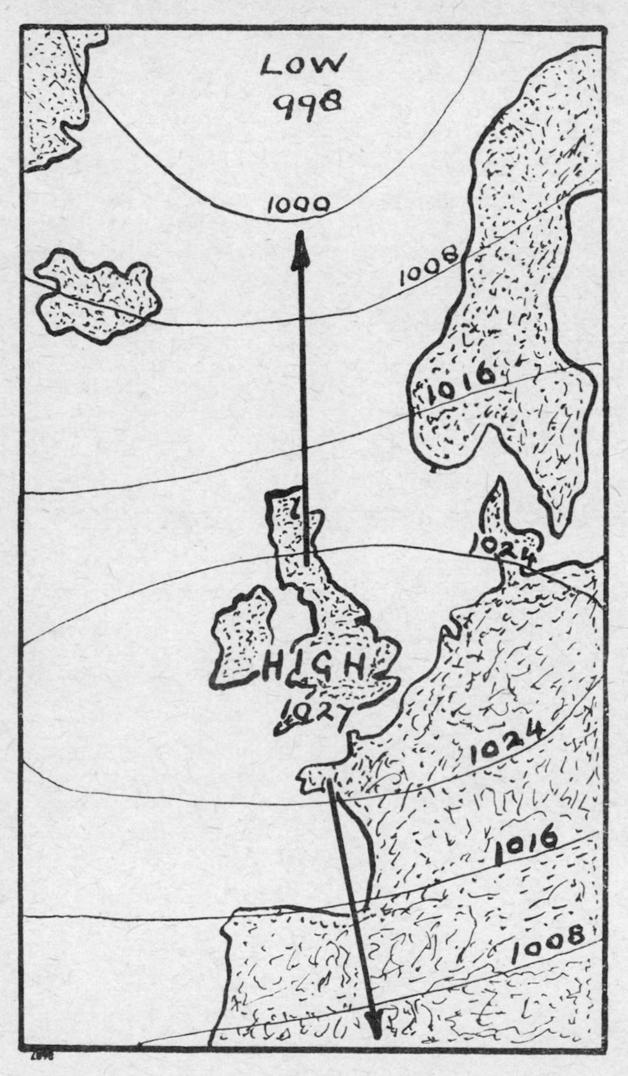

Fig. 7.

to the fact that the earth is not stationary but is rotating about its axis.

Geostrophic Force. Geostrophic force, which acts on any body which has movement of its own and is free of frictional connection with the earth's surface, has the effect of deflecting the direction of motion of the body to the right if it is in the Northern hemisphere and to the left if it is in the Southern hemisphere. It arises by reason of the fact that, since the earth is a sphere rotating Eastwards on its axis, different parts of the earth's

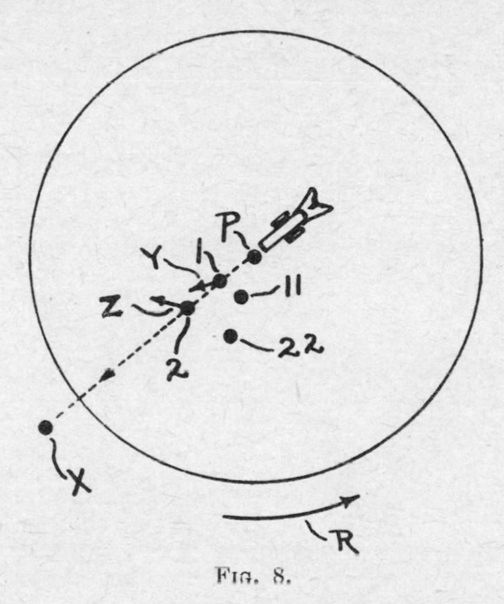

FIG. 8.

surface have different actual Eastward speeds depending on the latitude, parts in low latitudes having higher speeds than parts in higher latitudes, the speed at the equator being a maximum and the speeds at the two poles being zero. Consider the simplest case of a small area round the North Pole and, again for simplicity, regard this area as though it were a disc rotating Eastwards with the pole at the centre.

In Fig. 8, P is the North Pole and the arrow R represents the direction of rotation. Now suppose a shell to be fired from the pole towards a target X fixed in space. The actual velocity of the shell is therefore in the broken-line direction PX. However, by the time the shell has reached the point 1 in the line PX, the point on the disc originally under 1 will have moved to 11 and, viewed from this point on the disc, the shell will not appear to be moving in the direction PX but to the right of this direction—i.e. in the direction indicated by the arrow Y. Similarly, by the time the shell has reached point 2 in the line PX, the point on the disc originally underneath 2 will be at 22 and the apparent movement of the shell, as seen from 22, will be still further deflected to the right as indicated by the arrow Z . . . and so on until the maximum deflection of 90° is reached and the apparent direction of the shell is perpendicular and to the right of the original direction. It will at once be seen that if the disc had been representative of a small area round the South Pole, with that pole at the centre of rotation, the same sort of effect would have been produced except that the deflection would have been to the left instead of to the right. It may be shown that the same sort of effect is produced irrespective of the direction in which the shell is fired and the point from which it is fired, so that we may consider geostrophic force as a force which operates at right angles to the direction of motion of a body moving freely with respect to the earth's surface, to the right of that direction in the Northern hemisphere and to the left of it in the Southern hemisphere.

Geostrophic Wind. The air at 1500 ft. may be considered as free of frictional connection with the earth's surface, and air at this height is therefore, like the imaginary shell in Fig. 8, subject to geostrophic force acting to the right in the Northern hemisphere and to the left in the Southern. If, therefore, we have a steady wind blowing in a constant direction (i.e. along a great-circle track) with respect to the earth, it can only be because the geostrophic force is being opposed by and counterbalanced by some other force. This other force is the gradient wind force already referred to, and if these two forces balance they must be equal and acting in opposite directions in the same straight line. The stable wind condition at 1500 ft. must accordingly be as pictured in Figs. 9a and 9b for the Northern and Southern hemispheres respectively. In both cases the geostrophic and gradient wind forces are equal and opposite, both being at right angles to the wind direction. In the Northern hemisphere (Fig. 9a) the geostrophic force GE is to the right of the wind direction and the balancing gradient wind force GA is to the left. In the Southern hemisphere the opposite applies. In both cases the wind is along the isobars because the gradient force GA is at right angles to them, but the direction is opposite in the two hemispheres, with low pressure on the left in the Northern hemisphere and on the right in the Southern.

Wind Circulation. From the foregoing it follows that, with closed pressure systems—i.e. in systems where the isobars are closed on themselves—such as occur in practice, the winds at 1500 ft. will be along the curved isobars, and that in the Northern hemisphere they will circulate anti-clockwise round lows and clockwise round highs. In the Southern hemisphere the circulation directions are the opposite—i.e. clockwise round lows and anti-clockwise round highs. In both cases the surface winds,

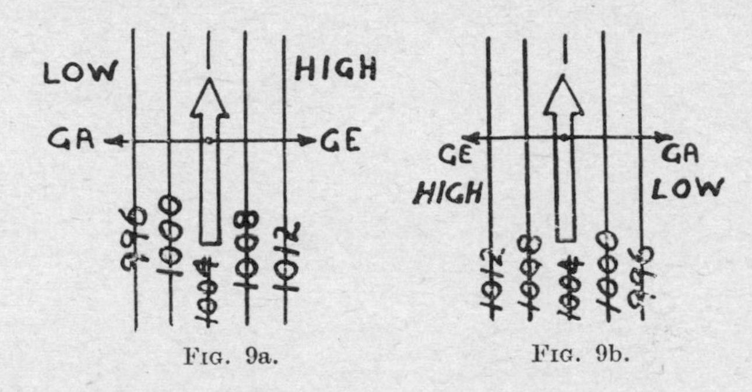

FIG. 9a. FIG. 9b.

because they are not free of the earth's surface but are, on the contrary, subject to the effect of friction with the earth, will tend somewhat in the direction of the gradient wind force and therefore be a little inclined off the isobars towards low pressure or away from high pressure. The amount of this inclination varies with circumstances, but, in general, over the sea, it is normally between one and two points of the compass. Figs. 10a, b, c and d typify the circulation directions around closed pressure systems —a and b being for the Northern hemisphere and c and d being for the Southern. In these figures the arrowheads on the isobars indicate the wind directions at 1500 ft. and the "wind arrows" indicate the surface wind directions.

Geostrophic Wind Scale. Since the gradient wind force *increases* with increase in gradient and geostrophic force *decreases* with

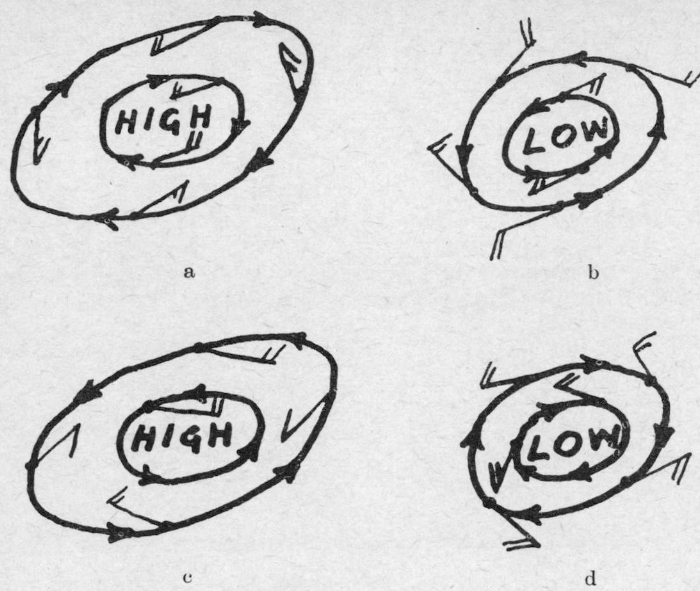

FIG. 10.

increase in latitude (because, of course, the Eastward speed of the earth's surface decreases as the poles are approached), and since, under stable conditions, the gradient and geostrophic wind forces are equal and opposite, it follows that (1) in any given latitude the geostrophic wind speed increases with increase of gradient, and (2) for any given gradient the geostrophic wind speed will depend on latitude, being greatest in low latitudes near the equator and least in high latitudes near the poles. Some sort of geostrophic wind scale is therefore essential if wind speeds are to be ascertained from a meteorological chart. Fig. 11 is an example of a geostrophic wind scale showing the connection between pressure gradient, latitude, and wind at 1500 ft.—i.e. geostrophic wind speed.

This scale is drawn for isobars at 2 mb. intervals and to use it the spacing distance, in sea miles, between two isobars differing by 2 mbs. is measured, the latitude is noted, and the intersection of the appropriate "spacing" line with the appropriate "latitude" line on the scale is observed. If this intersection falls on one of the curves, the speed marked on that curve gives the wind speed in knots at 1500 ft. If the intersection falls between two curves, the wind speed can be estimated accurately enough, for practical purposes, by interpolation by eye between the two curves in question. Thus, in latitude 60° a spacing of 300 miles between isobars differing by 2 mbs. gives a geostrophic wind speed of 5 knots. In the same latitude a separation of 200 miles gives a speed of about 8 knots. In many cases the projection used for a meteorological chart is such, and the difference of latitude covered by the chart is such, that it is possible to provide it with a simpler form of scale prepared for a particular latitude in the area charted and which can be used, without significant error, all over the chart. It is customary, in such cases, for the geostrophic wind scale to be prepared to suit the scale of the chart and to give geostrophic wind speeds directly, and many meteorological charts have such wind scales printed on them.

SPACING IN SEA MILES OF ISOBARS AT 2MB INTERVALS

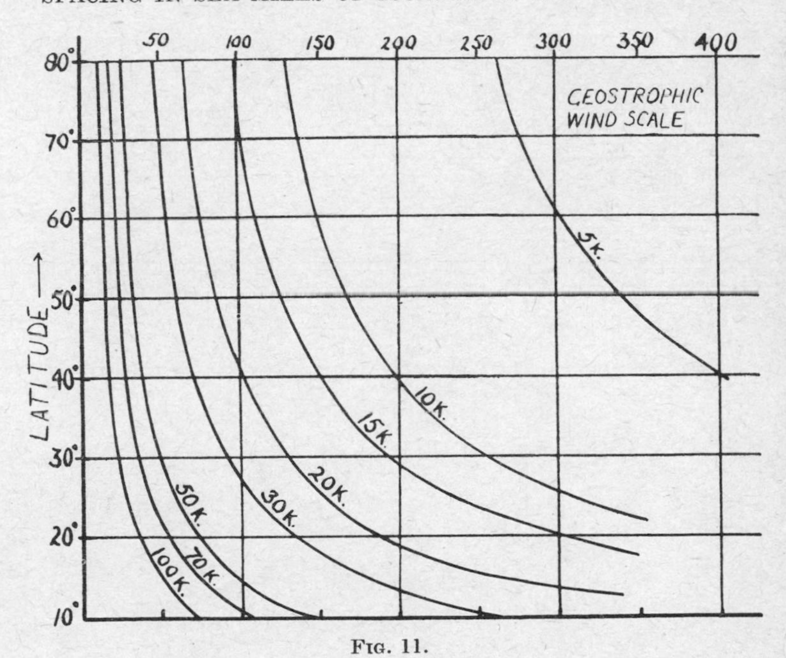

Fig. 11.

Fig. 12 shows a wind scale of this nature. Having drawn the isobars, one simply measures with the dividers the spacing between two isobars at the pressure interval for which the scale is prepared, transfers the dividers to the scale with one point on the left-hand end, and reads off the wind speed indicated by the other point of the dividers. Thus, in Fig. 12 the geostrophic wind speed being read off by the dividers is about 12 knots. It is important that the dividers be opened to the spacing between isobars at the interval for which the scale is prepared: thus, if the scale is for a 2 mb. interval, the dividers must be opened to the spacing between isobars differing by 2 mb. *The dividers must not be opened to the spacing between isobars differing by, say, 4 mb.,*

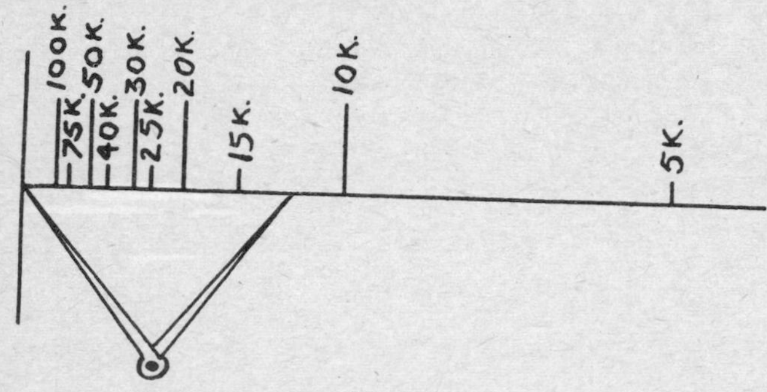

Fig. 12

the wind read off, and the reading multiplied by 2, because the scale is far from linear and therefore the result achieved by doing this will be badly wrong. If, as may happen, the isobars are drawn at 4 mb. intervals and the scale is for 2 mb. intervals, *divide (by eye) the separation between adjacent isobars by 2 and open the dividers to this value.*

Surface Winds. As has already been seen in Fig. 10, one effect of earth's friction is to cause the surface-wind directions—surface wind may be regarded as wind at a height of 30 ft.—to be inclined inwards towards low pressure and outwards from high pressure, in relation to geostrophic wind directions. Friction also, of course, reduces surface wind speeds with relation to geostrophic wind speeds. In the absence of shelter and special local variations, such as land and sea breezes (to be dealt with later) experienced

6. Cirro-Stratus —large Cumulus below. (Mediterranean: autumn)

7. Cirrus, plentiful and increasing. (Kattegat: summer)

8. Cirro-Cumulus: middle portion of picture, with large Cumulus below. (Bay of Biscay: winter)

9. Cirro-Stratus, plentiful and increasing: Cumulus below. (North Sea; spring)

near coasts, surface wind speeds over the sea may be taken, without material error, as being two-thirds of geostrophic wind speeds. Over land, where the friction is higher, surface wind speeds are usually about one-third of geostrophic wind speeds. The mariner will not go far wrong, therefore, if he uses the met. chart to find the geostrophic wind speed and then, under open sea conditions, estimates the wind speed he will experience as being two-thirds of the geostrophic speed and in a direction inclined towards low pressure by about one compass point with respect to the geostrophic wind direction. It is, indeed, quite practical to prepare a modified geostrophic wind scale to read open-sea wind speeds directly; and although the professional meteorologist never uses such a scale because it does not serve his needs, such a scale has obvious advantages for the more limited needs of the little-ship man, and later in this work a practical chart with such a wind scale on it will be provided.

CHAPTER V

AIR TEMPERATURE EFFECTS: LAPSE RATES AND STABILITY

Water Vapour. Cloud, fog and precipitation—fog is only cloud on the surface and precipitation is only a general term to include rain, drizzle, hail, snow and sleet—are all phenomena due, directly or indirectly, to the inability of air, under the conditions it is experiencing at the time, to retain water in the form of water vapour. Water vapour is a gas, and, being transparent and colourless, is invisible and indistinguishable by eye from air. Only when it condenses does it become visible as cloud or fog. When a kettle is boiling vigorously, producing a jet of visible steam, it will be observed that there is a short length of invisible jet close up against the spout and between the spout and the visible steam. This invisible part of the jet is water vapour: the visible steam is condensed water vapour. The amount of invisible water vapour which a given quantity of air can retain in this form depends upon the air temperature. The amount increases with air temperature at a faster rate than simple proportionality. Thus, if air at 30° F. is warmed to 50° F. the amount of water vapour it will hold is about doubled, and if it is warmed still further to 70° F. the amount is rather more than redoubled.

Relative Humidity. Dew Point. The *relative humidity* of air—often called, simply, its humidity—is the amount of water vapour it actually holds expressed as a percentage of the amount of water vapour it could hold at whatever its temperature is. Thus a cubic metre of air at 50° F. can retain approximately 10 grammes of water in the form of invisible water vapour. If it actually has 10 grammes of water its humidity is 100%: if it has only 7·5 grammes, the humidity is 75%. If air of less than 100% humidity is cooled down enough, its humidity rises until it reaches 100%. At this temperature the air is said to be saturated and the temperature in question is called the *dew point*. If air is cooled below its dew point—i.e. beyond the point at which 100% humidity is reached—the excess water vapour is condensed out and becomes visible as cloud, if the air is above the surface, or fog if at surface level.

Pressure and Temperature. Air can be warmed or cooled in a number of different ways, one obvious way being by blowing over a sea or land surface which is warmer or colder than the air itself. Another, and exceedingly important, way in which warming or cooling can occur is by a change in height. The air pressure at any place on the earth's surface is obviously due to the weight of the column of air above it; therefore if any particular "parcel" of air is moved upwards for any reason, the weight of air above it becomes less, the pressure on it is reduced, and it expands. Similarly, if it is moved downwards, the weight above it is increased, the pressure on it is increased, and it is compressed. It is a well-known principle of thermodynamics that if a gas (air is the gas considered here) which neither receives heat from nor gives up heat to any external body has its pressure and volume changed, its temperature changes also, increasing with increase of pressure and decreasing with decrease of pressure. This principle is utilised, for example, in the Diesel engine, in which compression of air in the cylinder raises its temperature to a value at which it will ignite oil fuel. Consider now Figs. 13a, b and c, which show a piston within a lagged cylinder supposed to be filled with completely dry air containing no water vapour.

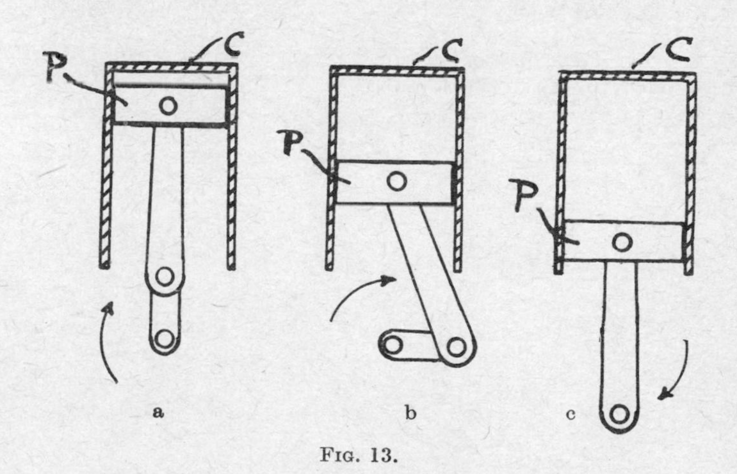

FIG. 13.

When the piston P is at the top of its stroke as in Fig. 13a, the air in the cylinder C is at maximum pressure and of minimum volume and its temperature is at a maximum. As the piston comes down to some intermediate position, such as that of Fig. 13b, the pressure decreases, the volume increases and the temperature drops, until finally, when the piston is at the bottom of its stroke as in Fig. 13c, the pressure and temperature are at a minimum and the volume is at maximum.

Dry Adiabatic Lapse Rate. If dry air, or air which never reaches saturation (100% humidity), is lifted in height it undergoes a reduction in pressure, expands, and behaves in the same way as does the air in the cylinder C of Fig. 13 when the piston P comes down. The rate at which its temperature will drop due to expansion as it rises is (nearly enough) 5·4°F. per 1000 ft. of rise. This sort of expansion or compression—occurring with the air neither receiving nor giving up heat to an external body—is called *adiabatic*, and the rate of temperature change which occurs when air less than saturated expands or contracts adiabatically, by rising or falling, is the dry adiabatic temperature change rate. As stated, it is about 5·4° F./1000 ft. Now, as is well known, if one rises in the air the temperature actually experienced may or may not change—though it usually does—and may either increase or decrease. If, however, as one rises, the temperature actually decreases at the rate of 5·4° F./1000 ft. the air is said to have the *dry adiabatic lapse rate*, the term "lapse rate" being

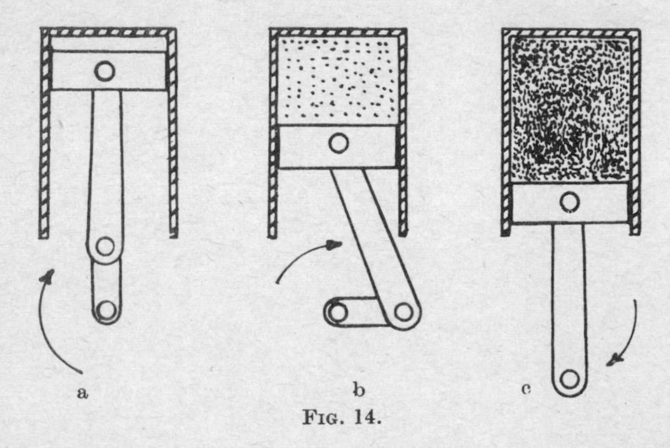

Fig. 14.

used for the actual rate at which temperature changes with increase in height.

Saturated Adiabatic Lapse Rate. Now consider Figs. 14a, b and c, which are like Fig. 13 except that the air in the cylinder is no longer dry air but air containing water vapour. Suppose the amount of water vapour to be such that the air is well below saturation when it is fully compressed and of maximum temperature as in Fig. 14a. Then, as the piston comes down, the temperature will fall at the dry adiabatic temperature change rate until 100% humidity is reached. Suppose this point to be reached at the position shown in Fig. 14b. Up to this position the action will be exactly as in Fig. 13. From this point onwards, however, further downward movement of the piston causes the air to be cooled below its dew point and it will begin to condense out its excess water vapour. Now, of course, it requires heat to make water change its state to water vapour, and when water vapour changes its state back again, as it does when condensing, this heat—the so-called latent heat—is given off again. Accordingly, as soon as expansion reaches the point where condensation starts, and for so long as it continues, heat is given up to the air by the condensing water vapour and there is a warming effect which offsets, to some extent, the cooling effect of expansion. The net result is that, as the piston continues to move down from the position shown in Fig. 14b to that shown in Fig. 14c, the temperature continues to fall but at a slower rate, which is called the saturated adiabatic temperature change rate. If saturated air is lifted in height it expands and cools; but because, during such lifting, water vapour is being condensed out just as it is in the cylinder between Figs. 14b and 14c, the cooling rate is slower than it would be were the air unsaturated. The saturated cooling rate averages about 3·2° F./1000 ft.—less than the unsaturated or dry rate. If, as one rises in the air, the temperature actually experienced falls at about 3·2° F./1000 ft., the lapse rate is said to be the *saturated adiabatic lapse rate*.

Cloud Formation. It will now be seen that if unsaturated air is lifted by any means it will expand, cool at the rate of 5·4° F./1000 ft. until it is saturated, and thereafter, on further lifting, cool at the rate of about 3·2° F./1000 ft. Conversely, if saturated air with excess water vapour (cloud) is driven downwards by any cause, it will warm up at the rate of

Fig. 15.

about 3·2° F./1000 ft. until it is only just saturated, and thereafter warm at the higher rate of 5·4° F./1000 ft. In either case the level at which it is just saturated with no excess water vapour—this is the *condensation level*—will be the level of the cloud base. Also, even if the air is not rising or falling in height, cloud will form in it at the level at which the temperature is just below the dew point. Thus in Fig. 15 the broken line CL represents the condensation level—i.e. the level at which the temperature of the air is such that the air can just, but only just, hold its water content in the form of water vapour. Above this level the temperature is lower than this and clouds form as indicated. It may happen frequently that there are two different air streams at different heights with different temperatures and

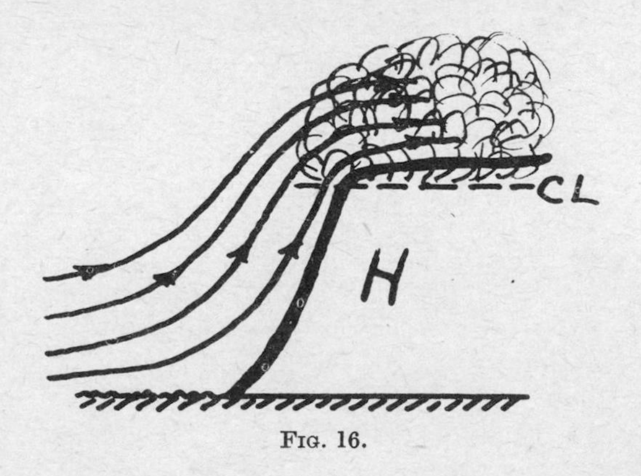

Fig. 16.

different moisture contents. In such cases there may well be two different condensation levels and two different cloud sheets, one above the other.

Fig. 16 represents a case in which unsaturated air, blowing in from the sea, encounters a high headland H and blows up over it. The consequent cooling produces saturation and condensation above the level CL and the headland is shrouded in cloud though there may be little or no cloud at sea. This case is very common and explains why lighthouses, originally sited on the tops of headlands, are sometimes abandoned in favour of new ones at the cliff foot. Fig. 17 shows the opposite case where a wind, cloud-laden over a high headland H and enshrouding the top of it,

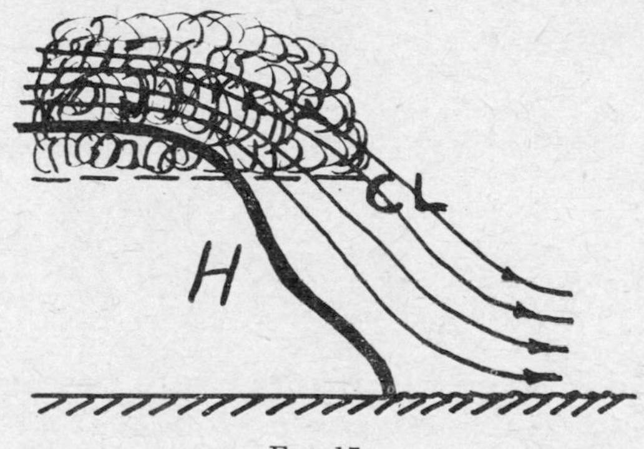

Fig. 17.

blows down towards sea level and is warmed in the process. When the level CL, at which the air can contain all its water vapour, is reached, the cloud disappears. This case is not as common as that of Fig. 16, but a world-famous example of it is given by Table Mountain in South Africa. In a South-Easterly wind the mountain-top is usually hidden in cloud which disappears (actually evaporates) quite sharply as it pours down over the edge so that it looks like a table-cloth and is, as a matter of fact, so called.

Stable and Unstable Air. It will be seen, from the foregoing, that we have two quite different aspects of air temperature to consider. The first, involved in the idea of lapse rate, is the temperature distribution vertically that is actually present—i.e.

the way in which a thermometer would change in reading if it was carried up vertically through the air. The reading might decrease at the dry adiabatic lapse rate or at the saturated adiabatic lapse rate, or at some other rate, or it might remain fairly constant, or it might, in some circumstances to be explained later, even increase. Different sorts and rates of changes could be present in different layers of air. The second aspect is involved in the idea of temperature changes produced, as explained in connection with Figs. 13 and 14, by movement upwards or downwards of a particular "parcel" of air. Consideration of these

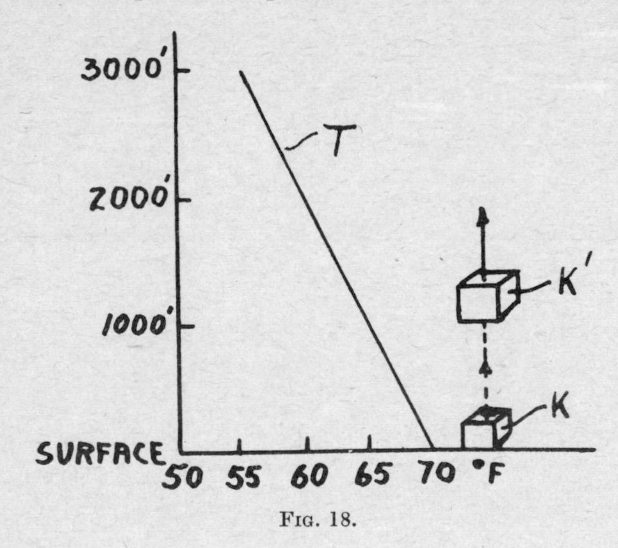

Fig. 18.

two aspects together leads to the notions of stability and instability, which are most important in deciding the weather characteristics of an air stream.

Suppose, as represented in Fig. 18, the air has a high lapse rate —for example, a lapse rate of about 5° F./1000 ft. In Fig. 18 this lapse rate is represented by a graph T connecting temperature on the bottom scale with height on the side scale. Now suppose that a small "parcel" of this air on the surface and represented by the cube K is started in an upward movement—for example, by blowing up the side of a hill. For the sake of simplicity of argument, suppose also that this parcel of air is just saturated when in its original position. Then, as it moves up it will cool at the saturated adiabatic rate of 3·2° F./1000 ft. By the time it reached 1000 ft. it would have cooled 3·2° F. below its original temperature when on the surface. But the air originally at 1000 ft. is, owing to the lapse rate, 5° F. below that on the surface. Accordingly when the lifted parcel has reached the position K′, 1000 ft. up, it will be $5 - 3{\cdot}2 = 1{\cdot}8$° F. warmer than the rest of the air at this level. Since warm air is lighter than cold, there will now be a force on it, due entirely to its having been raised, tending to make it rise still further. In other words, the conditions pictured in Fig. 18 are such that there is an inherent tendency for upward movement of air, and any upward movement, once started, will continue and proceed so long as the lapse rate is high enough to make air, cooled by expansion on rising, warmer and therefore lighter than the air already at the new level. *When these conditions obtain the air is said to be unstable.* Obviously the characteristic of unstable air is vertical movement. Clouds tend to be broken and of marked vertical development—in other words, of pronounced cumulus type; the greater the instability, and the greater the height of air over which it occurs, the greater the vertical development of clouds if there is enough water vapour present to cause them to form. Winds tend to be gusty and

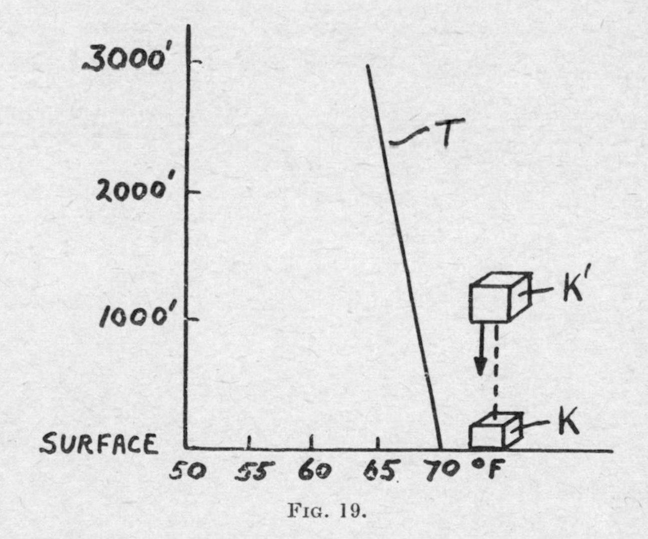

FIG. 19.

turbulent. The air is what an aviator calls "bumpy", and "bumpiness" due to mountains is felt at a great height above them. Visibility will tend to be good because smoke and small particles of dirt are carried up by the upward-moving air streams and soon dispersed. Funnel smoke goes sharply up and disappears quickly.

Fig. 19 shows, in the same manner as Fig. 18, the conditions for stable air. Here the lapse rate is small, being shown as only 2° F./1000 ft. Under these conditions, if the just saturated parcel K of air is raised 1000 ft. to K′ it will be cooled, as before, through 3·2° F., but it will now be in a layer of air whose temperature is only 2° F. below the air on the surface. It will accordingly be $3{\cdot}2 - 2 = 1{\cdot}2°$ F. colder than the air already at its new level and therefore heavier than that air. It will accordingly sink. When these conditions obtain—i.e. *when the cooling of air by expansion on lifting makes it colder and therefore heavier than the air already at the new level—the air is said to be stable.* In stable air vertical movement is automatically opposed, and if it is forced to occur, as by the wind blowing up a hillside, it does not go on longer than it must. Clouds, if there is enough water vapour to produce them, are stratiform rather than cumuliform and tend to be without marked vertical development and more or less unbroken. Winds, unless strong enough to be made gusty by surface irregularities, tend to be steady. The air is "smooth", not bumpy. Visibility tends to be only moderate or even poor because there is no vertical movement to disperse smoke and "dirt". Funnel smoke hangs about at a little more than funnel-top height. Summarising the main characteristics of stable and unstable air in tabular form, we have:

Unstable Air	*Stable Air*
Cumulus type clouds	Stratus type clouds
Gusty winds	Steady winds unless very strong
"Bumpy" air	"Smooth" air
Good visibility	Moderate or poor visibility

CHAPTER VI

RAIN, SNOW, HAIL AND FOG

Rain and Drizzle. The minute particles of water into which water vapour condenses to form cloud are small and light enough to float in the air. If, however, there is any great internal movement, and particularly upward movement, in the cloud they will coalesce on the tiny nuclei of salt and other impurities always present in the atmosphere and form separate droplets. Upward movement of these droplets cools them further, producing more condensation and coalescence, and so the droplets grow in size until they are heavy enough to fall through the air. *Rain*, according to the dictionary, is merely condensed moisture of the atmosphere falling in separate drops and, strictly speaking, includes *drizzle*. It is convenient, however, to use the two terms rain and drizzle, but the distinction between them is only one of *size* of drop (*not* number of drops), drizzle being merely rain in which the average diameter of the drops is less than a certain amount, which is more or less arbitrarily chosen at $\frac{1}{50}$ of an inch. The boundary between rain and drizzle is thus by no means sharp, the one merging almost imperceptibly into the other. Because drizzle is only very fine rain it is often in practice difficult to distinguish it from fog.

The maximum speed at which a raindrop can fall through the air is determined by its size, the speed attainable by a large drop being greater than that attainable by a small one. If, therefore, there is an up-current of air, the only raindrops which can reach the ground through it are those big enough to reach a falling speed exceeding the upward speed of the up-current. It follows, therefore, that heavy rain composed of large drops is associated with unstable air, and air having a large temperature lapse rate, for it is in such air that up-currents of high speed are most likely to occur. One of the principal reasons why tropical rain is usually so much heavier than rain in the temperate latitudes is that the high temperatures produced on land by solar heating in the tropics often give a very high lapse rate in the lower part of the atmosphere with consequent powerful up-currents, so that only very large raindrops can fall fast enough to reach the earth at all.

Rain "Shadows". It is because the size of the raindrops which can reach the earth is determined by the upward speeds of up-currents through which they must fall that high ground, such as headlands, very commonly throw well-marked "rain shadows" on their lee sides, often giving much lighter rain on the lee side than on the weather side—or even no rain at all.

Consider Fig. 20, which shows the weather and lee sides of a headland H with the wind blowing as indicated by the arrow-headed lines. On the weather side the raindrops must be big

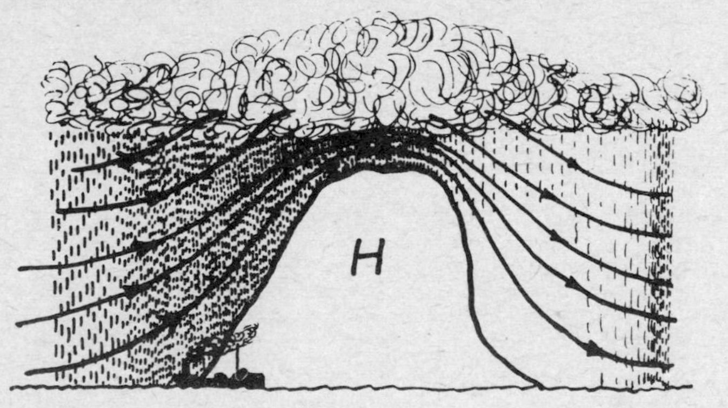

FIG. 20.

enough to fall through the upward-moving air stream to windward of the head, but on the lee side the air has a downward component and accordingly considerably lighter rain may be expected. Indeed, this effect is not infrequently so well marked that there may be no rain on the lee side at all. In general the rain "shelter" given by headlands in this way extends pretty well as far as, and sometimes further than, the protection given by the head from the wind itself.

Cloud-bursts. There is a limit to the speed at which a raindrop can fall in relation to the air in which it is moving; for when a limiting speed is reached, a raindrop big enough to attain that speed gets broken up into small droplets by the air through which it is moving. Since small droplets cannot attain the falling speed of large ones, it follows that rain cannot fall at all if there is an up-current of more than a certain vertical speed. According to one generally accepted authority, this limiting upward speed is about 18 m.p.h. There is every reason to believe that up-currents exceeding this vertical speed are fairly frequent, especially in the tropics. When up-currents of such high speeds occur, the rain-drops, which, in effect, are held up by them, can increase greatly in size, and if, for any reason, the up-current ceases or falls off much in speed, or if the rain-bearing cloud passes out of the always more or less limited area where such an up-current exists, there is a sudden torrential downpour of very large drops—in short, a cloud-burst.

Types of Rainfall. As will have been seen, rainfall is a by-product of the lifting of air which contains water vapour, the lifting resulting in cooling and consequent condensation of the water vapour into the cloud from which the rain falls. Any process which lifts air is therefore capable of producing rain. The three main processes are convection, orographic action, and frontal action. Broadly speaking, therefore, there are three types of rain: namely, convectional, orographic, and frontal.

Convectional Rain Showers. Convection is the name given to circulation in a vertical plane brought about by temperature differences. It is the sort of thing which occurs when water is heated in a saucepan as shown in Fig. 21a, the nearest meteorological parallel to this being what occurs over a tropical island as represented (with a deal of artist's licence) in Fig. 21b.

In Fig. 21a the water immediately over the central flame at the bottom of the saucepan is heated, expands, and rises centrally,

FIG. 21a

FIG. 21b

flowing out towards the cooler sides of the saucepan where it descends, moves inwards and again comes over the flame. There is thus a circulation in the vertical plane as indicated by the arrowheaded dotted lines. Similarly, the tropical island of Fig. 21b, strongly heated by the sun, produces an up-current of air which, after it has risen as high as the heating at its base will drive it, divides outwards and descends over the sea, coming in to the island again as a cooled sea breeze. If the air has sufficient water vapour in it—and the air over an island is, of course, very apt to contain considerable water vapour—cumulus cloud forms with its base at the condensation level and, if the conditions are otherwise right, rain will fall. In warm climates, and in temperate climates in summer, land, and in particular islands, can often be detected from seaward while still well below the horizon, by cumulus which forms over them even when the sky over the sea is cloudless, or nearly so.

The tropical island case of Fig. 21b is, perhaps, the simplest case to understand and for that reason has been put first, but, of course, it is by no means the only case, or even the commonest case, of convectional rain. It is liable to occur whenever sufficiently humid air passes over heated land and becomes markedly unstable by heating over its lower layers. The land does not have to be warm in the absolute sense but only relatively warm in relation to the air blowing over it. Thus convectional rain, usually in the form of showers, is common in North European latitudes when cold North winds of unstable air blow in over land which, around mid-day or early afternoon, has become considerably less cold than the air above it. A *shower*, of course, is only localised rain falling from more or less detached clouds—i.e. clouds of the cumulus type—and showers are therefore associated with cumulus clouds of large vertical development. The rain accompanying a thunderstorm is the most striking case of convectional rain. A humid air stream coming in over strongly heated land can produce towering thunder-cloud, going up to 50,000 ft. or more, for the heated land, so much warmer than the air above it, produces a high lapse rate in the lower atmosphere and sets up very powerful convection currents—powerful enough to be dangerous to aircraft.

The severe revolving storms experienced in tropical waters, mostly in summer, can also produce powerful up-currents with consequent heavy convectional rain.

Orographic Rain. This type of rain is, as the name implies, caused by the earth's surface itself, being due to mountains or high tablelands rising up more or less sharply from the general level of the land or sea. With this type of rain the up-currents which, fundamentally, produce the rain are caused by the wind blowing against the windward side of the high land and being forced to rise to pass over it. Much of the rain which falls along the West coast of Scotland and the North-West coast of Wales is orographic, being caused by the humid Westerly winds from the Atlantic having to rise to pass over the mountains along the West Scottish coast and in N.W. Wales respectively. More extreme cases of orographic rain are to be found in the Pyrenees, where, at certain times of the year, thundery activity is the rule rather than the exception, and along the Western Ghats (mountains) on the West side of India during the South-West monsoon. This last case produces what is probably some of the heaviest rain in the world, for the S.W. monsoon wind comes in, warm and very humid, off the Indian Ocean and is then forced up to three or four thousand feet to pass over the Ghats.

Frontal Rain. *Fronts* will be explained more fully in the chapter on depressions. It is sufficient to say here that when two air streams of different temperatures meet, the warmer air, being

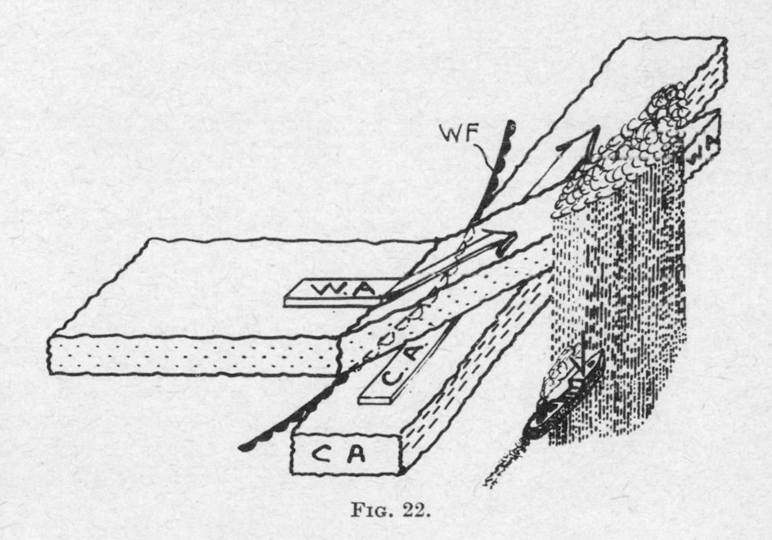

FIG. 22.

lighter than the colder, slides up over it along an inclined plane, each stream retaining its own identity and there being a quite definite boundary between them. There are various different sorts of fronts, but in all of them the warmer of the two streams rides up over the colder and is therefore lifted and cooled, producing cloud and, if the conditions are otherwise right, rain also. The action, so far as cloud and rain formation are concerned, is rather like orographic action, with the colder air stream playing the part of an inclined mountain face.

Fig. 22 is a diagrammatic representation of the sort of thing which occurs at fronts, the type of front chosen for exemplification being a warm front. Here a warm humid air stream WA, represented as a Westerly air stream, rides up over a colder air stream CA shown as a S.S.W. stream. The boundary line on the surface is the line WF which is drawn in the conventional manner adopted for the representation of warm fronts—i.e. as a line with rounded bumps on that side of it towards which the front is moving. In Fig. 22 the front is supposed to be moving in an approximately Easterly direction. As the warm air slides up over the cold it condenses to form cloud as indicated and rain is shown as falling from it. The ship in the picture is in cold air and is experiencing a S.S.W. wind. Above it, however, is a warm air stream blowing from the West and the cloud in that stream will, of course, be moving to the East. If there had been cloud in the cold air also and it had been broken enough to enable the higher cloud to be seen through it, the ship would observe two cloud layers at different heights moving in different directions, and this "crossing" of clouds at different heights, if observed, is a good indication of an approaching front. The column of air over her is made up of cold air with warm air on top of it. Had she been to the East of her illustrated position there would have been more cold air and less warm air over her. Similarly, had she been to the West of the illustrated position the proportion of warm air over her would have been still more, and had she been behind the front she would have had only warm air above her.

Snow and Sleet. The condition determining whether rain-clouds or snow-clouds will be found is only the value of the temperature at which condensation of water vapour occurs. If water vapour condenses at a temperature below 32° F., the freezing point of water, it condenses into crystals of the familiar feathery form known as *snow*. Any of the above described methods by which rain-clouds can be formed will form snow-clouds if the temperature is low enough. As snow falls it may be, and often is, warmed as it falls and may become rain on the way down, and there is no doubt that a great deal of rain on the surface starts falling as snow. Whether it remains snow by the time it reaches the surface depends on the surface air temperature. In general, if the surface air temperature is 37° F. or less, precipitated snow will remain snow at the end of its downward journey. The determination, in a forecast, as to whether snow or rain will occur at surface level is a very difficult matter and one which not infrequently falsifies the unfortunate meteorologists' prophecies, especially over land, because land can change in temperature quite rapidly. A forecaster who foretells rain on the basis of an estimated land temperature 6 hours hence of, say, 39° F. has only to be two or three degrees high in his estimate to get snow instead. At sea, fortunately, things are a good deal easier because sea retains the same temperature far longer than land. *Sleet* is only a half-way house between rain and snow—wet snow, or rain and snow mixed—and can be produced by any of the methods which will produce rain or snow.

Hail. *Hail*—ice pellets—is a phenomenon associated with extremely powerful up-currents of air and therefore with great air instability and cumuliform clouds of large vertical development, in particular the towering thunder-cloud. Raindrops form by condensation in the lower parts of such a cloud, but the up-current is so fast as to carry them up. When they get high enough for the temperature to be below freezing point they become what is termed "super-cooled"—i.e. they are still water drops but are below the normal freezing point. (There is good reason to believe that water drops can exist as such below 32° F.) As they get still higher they reach a region in the cloud where condensation of water vapour to ice takes place, producing an ice covering which grows on them until they become heavy enough to fall. When they do, they again reach a region where there are super-cooled water drops, and water from these drops freezes on to them as ice. They may then fall to earth or may encounter an up-current powerful enough to carry them up again so that the process is repeated—perhaps several times—until they are heavy enough to fall clear. The more violent the up-currents and the more towering the thunder-cloud, the larger will be the hailstones which finally do fall. In the tropics, where thunder-clouds of immense vertical development are not uncommon, hailstones of great size are experienced—hailstones weighing over 2 lb. and the size of a large orange have been recorded.

Frontal Fog. Fronts, and especially warm fronts like that pictured in Fig. 22, often produce fog ahead of them. The cloud just ahead of an active warm front is often very low—it requires very little lifting of warm, humid air, already nearly saturated, to become completely saturated and produce thick cloud and drizzle or rain—and this drizzle or rain may cool the air through which it falls and thus, by causing condensation from that air, produce fog ahead of the front. High winds, of course, tend to disperse fog, and therefore frontal fog, which moves with and at the speed of the warm air, is generally associated with light or light to moderate warm air winds and slow-moving fronts. Its production requires that the warm air shall be very humid and stable and that the cold air shall also be of considerable humidity and, if unstable, shall not have much instability in its lower levels. Frontal fog occurs in the form of a belt close ahead of the front. The belt is not very wide—it is seldom more than 30 or 40 miles wide—and, since it clears with the passage of the front, it is not very persistent.

Sea Fog. This is the most persistent and probably the commonest of all forms of fog. It is caused when humid air, already not far off saturation, blows over a sea surface cold enough to cool the lower levels of the air below the dew point—i.e. the condensation temperature. Again it requires for its formation that the air should be stable and that the wind should be light or light to moderate—between Force 1 or 2 and Force 4. Once it forms over sea of a given temperature it will persist until the air stream changes or the wind gets up enough to blow it away. A notorious area of persistent sea fog is the Newfoundland Banks, where the icily cold Labrador current is the cause. It is, however, not uncommon in British and Northern European waters, and will be produced by a "muggy" warm air stream coming in with a light wind over sea which is considerably cooler than the air.

Local Fog Effects. Localised but very troublesome sea fog often occurs in and off the mouths of rivers because river water, flowing out on top of the sea, may be cold enough to give the required air-cooling to produce sea fog even though the sea temperature proper is not.

A second local effect, which can be very important to the mariner making a landfall, is caused by land and sea breezes. On clear days and nights land and sea breezes (these will be dealt with more fully later in this book) occur along coastlines, an onshore breeze occurring in the late morning or early afternoon when the land has warmed up, and an off-shore breeze occurring some hours after sundown when the land has cooled off again. These breezes are purely local and extend only a few miles on either side of the coast. When there is sea fog, the land breeze at night will usually drive it and keep it off the coast so that a harbour is clear. When the sea breeze sets in, it brings the fog in with it. A harbour which is absolutely clear may change to nil visibility in half an hour during an afternoon and may clear as quickly at night.

A third but less important local effect is purely tidal and occurs in estuaries with considerable areas of banks which dry at low tide—e.g. the Schelde and Thames estuaries. When the banks are exposed around low water they may get warmed enough to cause convection (up-currents) or warm the air above the condensation level and thus dissipate the fog locally or at any rate thin it and make it patchy. The effect is not well marked or to be relied upon, but it does occur and is worth noting.

Land Fog. This occurs over land, mainly low-lying land, at night or in the early morning if the sky is clear enough to allow the land to cool off to below the condensation temperature of warmer humid air above it. Again it requires that the air shall be stable and the wind only light. It is of interest to the mariner only in so far as it may obscure harbours or navigation aids on low-lying coasts or, under the influence of an off-shore breeze, blow out to sea. It never extends far to sea.

Sea Smoke. This is a very low-lying, thin sort of fog and its appearance is best described in the statement that it looks as if the sea were steaming. In a sense this is what is happening, for it is actually caused by intense evaporation of water from the sea. It is largely a coastal phenomenon and is caused by a land breeze which is much colder than the sea—30° F. or so colder than the sea—blowing out from the land. It is naturally mainly a cold polar phenomenon, but it is fairly common along the Norwegian coast, is not infrequently experienced on Baltic and Danish coasts and is not unknown on Scottish coasts.

Haze and "Dirt" Fog. Haze and "dirt" fog are the two stages of intensity of the same thing—particles of smoke, dust or other matter suspended in the air. They can occur only in light winds and with stable air. An extreme case of dust haze, sometimes too thick, perhaps, to be properly called haze, is the famous "Harmattan" or "West Coast Doctor"—a dust-laden N. or N.E. wind which blows off the Sahara Desert and a few miles out to sea across the Atlantic seaboard of N.W. Africa. In general, however, haze does not much matter at sea except, perhaps, to the navigator, whom it will often rob of a sight by blurring his horizon. Haze thick enough to matter to the mariner—i.e. "dirt" fog—is never experienced any great distance from land, but it can be a great nuisance by blowing over harbours, estuaries and coasts which are only a few miles down-wind of large smoky industrial areas. Thus, a London fog blown by a light Westerly wind over the Thames Estuary can bring some shipping to a standstill.

CHAPTER VII

THE MECHANISM OF THE DEPRESSION: FRONTS

As has already been pointed out, there are two general classes of causes responsible for weather phenomena: namely, causes due to the properties of air masses moving over land or sea, and causes due not so much to the properties of air masses themselves as to the physical changes which occur when one air mass substitutes itself for another. Weather due to the former class of cause may be called "air mass weather"; that due to the latter class of cause may be called "frontal weather". Frontal weather is a dominant feature of weather conditions in temperate latitudes such as our own, particularly in depressions—i.e. moving centres of low pressure.

Formation of a Frontal Depression. Imagine two air streams, one of which is warm and humid—for example, of tropical origin—and the other of which is cold and drier—for example, of polar origin—to be flowing alongside one another in opposite directions as shown in Figs. 23a and 23b, of which the former is a diagrammatic perspective view and the latter is a plan view looking down.

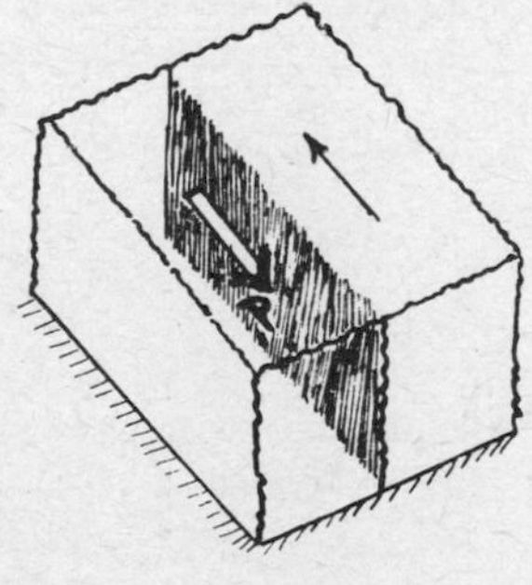

Fig. 23a.

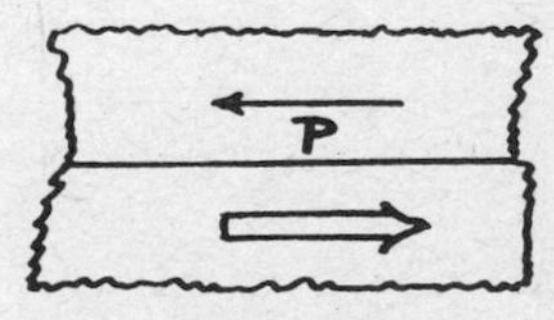

Fig. 23b.

Suppose that initially the common boundary between the two masses—this boundary is indicated by shading in Fig. 23a and by a line in Fig. 23b—is a vertical plane. Obviously this condition is not likely to last for long, but it can exist. The directions of movement of the air in the two streams are parallel but opposite, that in the warm air stream being indicated by the wide double-line arrow and that in the cold stream by the simple single-line arrow. Let the barometric pressure at some place at sea level in the cold air and near the common boundary be of some value p.

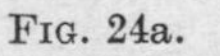

Fig. 24a.

Fig. 24b.

Now suppose that, as a result of some disturbance or of a small change in either wind direction, the warm air pushes out a bulge into some part of the volume previously occupied by cold air, of course climbing over the cold air since it is lighter than cold air. This is pictured in Figs. 24a and 24b, where again the boundary, now bulged, is indicated by shading and the direction of warm and cold air by arrows as in Figs. 23a and 23b. The barometric pressure under the bulge will now be reduced to a new, lower value p because part of the air above it is warm, whereas previously it was all cold. In other words, there is now a "low" at p. Assuming the Northern hemisphere, there will now be forces in play tending to bend the wind directions anti-clockwise as indicated by the arrows in Figs. 24a and 24b because, as was explained in Chapter IV, a low pressure in the Northern hemisphere produces an anti-clockwise circulation around itself. The new, changed directions of air motion are such as to cause the bulge, originally small, to grow as the warm air pushes out at an inclined angle into the space previously occupied by cold air. Also the bulge itself will move along the boundary wall something after the manner of a wave, the advancing warm air blown by the warm air wind climbing over the cold air in front of the leading side of the bulge and the cold air behind the moving bulge cutting in under the warm air along the trailing side of the bulge.

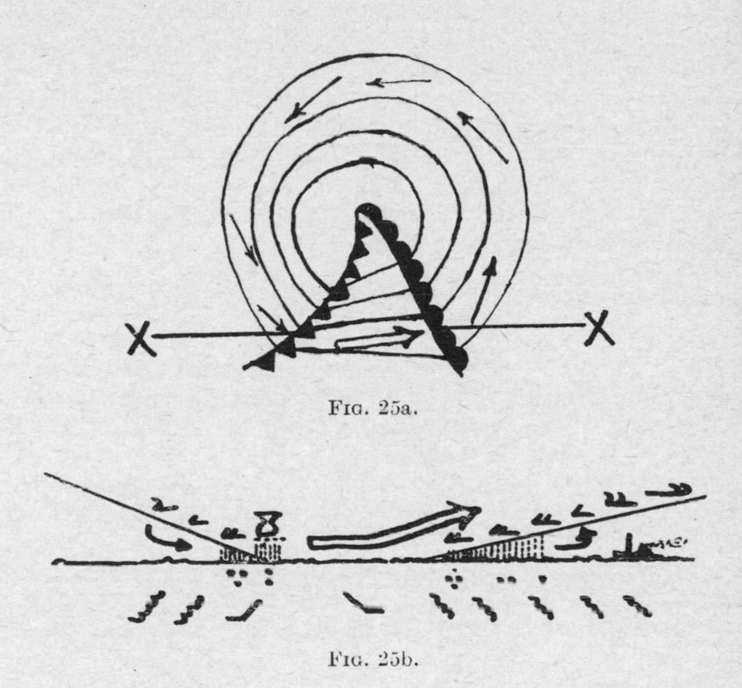

Fig. 25a.

Fig. 25b.

Warm Sector. The final result will be like that shown in Fig. 25a, which is a plan view, like those of Figs. 23b and 24b, but drawn for sea level, showing the now well-developed bulge and indicating its leading and trailing sides in the conventional manner adopted by meteorologists. Fig. 25a also shows, besides the warm and cold air wind arrows, the isobars representing the pressure distribution which has been set up. The picture is that of a developed frontal "low". That part of it where warm air is present on the surface—i.e. that part of it where the broad warm wind arrows are shown—is the *warm sector*.

Warm and Cold Fronts. That part of the boundary of the warm sector where warm air is replacing cold on the surface—i.e. the leading side of the bulge—is a *warm front* and is indicated in Fig. 25a by the standard conventional symbol consisting of a line with rounded projections on the side towards which it is moving. That part of the boundary of the warm sector where cold air is replacing warm on the surface—i.e. the trailing side of the bulge—is a *cold front* and is indicated in Fig. 25a by the standard conventional symbol consisting of a line with triangular projections on the side towards which it is moving. Fig. 25b is a cross-section taken on the line X–X of Fig. 25a, showing the two fronts as inclined boundary surfaces with warm air climbing over cold at and ahead of the cold front on the surface and cold air cutting in under warm at and behind the cold front on the surface.

Warm Front Weather. Imagine a ship to be as shown under the warm front of Fig. 25b and the whole system to pass over it from left to right in the figure. Suppose also that the pressure distribution over the whole depression remains unchanged while this is going on. Then the ship will experience a steadily falling barometer as the warm front on the surface gets nearer and nearer because, during this time, the air over the ship becomes more and more warm air and less and less cold air. The slope of the warm front varies, of course, from example to example, but a common value is about 1 in 100. As the warm air climbs up the frontal inclined surface it is progressively cooled by lifting and suffers condensation. High up on the inclined surface, generally about 5 miles up, cirrus cloud (ice crystal cloud) will form, and such cloud—the "mare's-tails" of the seaman—is an early sign of an approaching warm front. In an average case cirrus may be expected to appear more or less simultaneously with the beginning of a fall in the barometer at around four or five hundred miles ahead of the front on the surface. The ship will see the cirrus increase in amount and be gradually replaced by cirro-stratus, which is a high, thin, shapeless veil of grey through which the edge of sun or moon appears sharply defined. This growth of cirro-stratus accompanied by a falling barometer is the forecasting combination lying behind the old (and good) couplet:

"A greying sky and a falling glass
Soundly sleeps the careless ass."

Because the condensation causing the cloud formation is due to the lifting of a great mass of air as it rises up the inclined plane of the frontal surface, the cloud continues to be of the shapeless stratiform type and the ship will see lower and lower stratus cloud becoming thicker and thicker over her. The next stage after cirro-stratus is alto-stratus (watery sky), thickening, lowering and darkening, with, at some stage, rain (or if it is cold enough, snow) falling, usually intermittently at first. As the cloud lowers it becomes low stratus—this differs from alto-stratus only in being lower—with rain (or snow) continuing to fall until the front on the surface goes through. All this time, before the front on the surface passes, the barometer falls and, especially in the latter part of the period, the wind increases in strength because, in this part of a depression, the pressure gradient usually increases. Often there is a slight, temporary back of wind a short time before the front goes through and fog is not infrequent a short way ahead of the front. (A back of wind is, of course, a change in direction anti-clockwise as from S. to S.E. and a veer is a change in the opposite direction—e.g. from S. to S.W.) With a well-marked warm front the width of the rain belt in front of it is often from 100 to 200 miles and headlands and other high ground can be expected to be covered in stratus cloud well ahead of the front even if there is no surface fog. The commonly occurring temporary backing of the wind close ahead of the front underlies the couplet:

"If the wind shifts against the sun
Trust it not, for back it will run."

Passage of the Warm Front. When the warm front on the surface has passed there is a more or less sharp veer of wind—e.g. from about S. to about S.W.—for the ship is now in the warm sector and experiences the warm air wind. Also the barometer ceases to fall because now the air overhead is all warm and there is no longer any replacement of cold air overhead by warm. The general weather will be that appropriate to the air mass in the warm sector—commonly "muggy" with stable, humid air, only moderate visibility, and drizzle. (A later chapter deals with air-mass weathers.) The winds will be strong, and if the depression is deep enough, and the ship is near enough to the centre of the low, very strong winds or gales will be experienced. The combination of rain ahead of the front with strong winds behind it is recorded in the ancient couplet:

"*When the rain's before the wind,*
Then your topsail halyards mind"

which I rather prefer in a revised and more modern form, thus:

"*Rain before veer*
Hard blow is here."

Cold Front. Eventually the cold front passes and warm air overhead begins to be replaced by the cold air cutting in under it. The slope of the inclined frontal surface thus produced is a good deal steeper than the corresponding slope of the warm frontal surface—usually about twice as steep, 1 in 50 being a fair average figure. The same sort of weather phenomena occurs after the passage of the cold front as occurs before the passage of the warm front but in the reverse order and, because the steepness is so much more, with a good deal more vigour. Hence the old couplet:

"*When the wind's before the rain*
Soon you may set sail again"

or, in the revised version I prefer:

"*Veer before rain*
Set sail again."

Because of the rapid upward movement of the warm air along the steeply inclined cold frontal surface the rain behind a cold front is usually considerably heavier than that given by a warm front, but, for the same reason, the width of the rain belt is usually very much less. The upward movement of the warm air is, in many cases, violent enough to give rise to thunderstorms or line squalls on passage of the cold front and, frequently, to some rain a little ahead of it. A veer of wind to the wind direction in the cold air behind the front takes place and this veer is usually sharper and the wind greater than at a warm front. The barometer also rises sharply because cold air is replacing warm overhead and, because the isobars tend to be squeezed together (see Fig. 25a) behind the front, the wind may become even stronger and only afterwards gradually ease off. Hence the couplet:

> "*Sharp rise after low*
> *Oft foretells a stronger blow.*"

In general the rain (or snow) belt behind a cold front is of fairly short duration and soon gives way to the weather appropriate to the new cold air mass—commonly showery, with broken cumulus clouds and good visibility outside showers. A vigorous cold front must be treated with respect by the little ship under sail, because the sharp change in wind direction, the sudden gustiness and squalliness—as stated above, there may be a line squall from a well-developed cold front—the confusion of sea which naturally accompanies a sharp change in direction of a strong wind, the possible increase of wind behind the front and the fact that the wind has "weight", as a sailor says (a good expression, this, the air actually *is* heavy), can give the little-ship master plenty to think about if he has not shortened sail in time. Once, however, the rain belt has gone through and the wind has eased, weather conditions are likely to improve until the next fall of the barometer heralds the coming of a new depression. Hence the couplet:

> "*Only when the glass is rising*
> *Takes his watch below the wise 'un.*"

The principal weather and barometer phenomena experienced are indicated in the appropriate places in Fig. 25b by the symbols recommended in Chapters II and III, to which reference should be made if they have been forgotten!

Occlusion. It will be obvious that in a frontal depression the speed of the warm front is governed by the warm air wind speed behind it, and similarly the speed of the cold front is governed by the cold air wind speed behind it. Just as the warm air, in order to advance at a warm front over the cold air ahead of it, must be moving faster than that air, so the cold air, in order to advance and cut in at a cold front under the warm air in front of it, must be moving still faster. Accordingly the cold front moves faster than the warm front and catches it up, at first near the centre of the depression and then gradually more and more along the length of the warm front.

Where a cold front has caught up a warm front on the surface, the part of the front where this has occurred is called an "occlusion", and a depression in which this has happened is said to be partly or wholly *occluded*—partly, when the front is occluded over only part of its length; and wholly, when it is occluded over its whole length. The standard symbol for an occlusion is a line with alternate rounded and triangular projections on the side towards which movement is taking place. Fig. 26a shows a

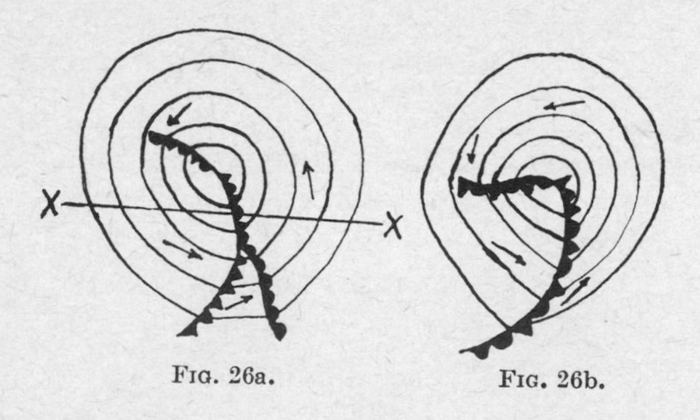

Fig. 26a. Fig. 26b.

partly occluded depression with an occlusion extending to the point where the front divides into warm and cold. Fig. 26b shows a wholly occluded depression. In both cases the warm air has been lifted right off the surface at the occlusion. In general the weather sequences associated with an occlusion are, as one would expect, a combination of the warm- and cold-front weather sequences, the latter following the former immediately. The veer of wind at an occlusion is, for obvious reasons, much greater than that normally experienced at either a warm or a cold front, and can easily be as much as 180°. Although the air behind an occlusion may be warmer, or of the same temperature as (this is very unusual), or colder than the air ahead of it, the last possibility is the commonest and an occlusion where this occurs is a *cold occlusion*. It has, as a rule, pronounced cold-front weather

Fig. 27a.

Fig. 27b.

characteristics behind it and a vigorous cold occlusion needs to be treated with all and more than the respect accorded to a vigorous cold front because of this and because of the magnitude of the veer. A *warm occlusion*, which is one in which the air behind the front is warmer than the air ahead of it, gives generally similar weather to a cold occlusion except that there is not the same vigorous upsurge of air behind the front and not the same gustiness and general liveliness when the veer of wind occurs. Fig. 27a is a cross-section at some such place as along the line X–X of Fig. 26a showing the structure of a cold occlusion, while Fig. 27b is a similar cross-section illustrating a warm occlusion. It will be noted that in Figs. 26a and 26b the occlusion is shown extending through and back beyond the centre of the low. This is quite common. Both parts of the occlusion—that above the centre of the depression and that below it—will swing anti-clockwise in accordance with the general air circulation round the low. That is why, in Figs. 26a and 26b, the projections on the line representing the front are shown on the right on the part of the line below the centre and on the left on the part above the centre. The great majority of depressions reaching the British Isles and Europe from the West are already partly occluded by the time they reach the land.

Summary of Frontal Phenomena. We may now set out, in a table, the phenomena which occur, in the order of occurrence, when the different sorts of fronts approach and pass, and by watching for these phenomena the seaman can do a good deal of quite practical short-range single-observer forecasting even if he has no official forecast to help him.

Approach of Warm Front or Occlusion

1. Increasing mare's-tails (cirrus). Barometer starts to fall.
2. Grey veil (cirro-stratus) spreads over sky. Barometer still falling.
3. Watery sky (thin alto-stratus) increasing and thickening. Intermittent rain or snow. Barometer continues to fall.
4. Stratus cloud thickens and lowers. Rain or snow becomes continuous. Wind usually begins to increase. Barometer still falling.
5. Very low cloud covering high ground. Continuous rain or snow. Sometimes fog. Often slight back of wind, which now usually increases rapidly. Sea rises. Barometer still falling.

Passage of Front

IF A WARM FRONT	IF AN OCCLUSION
1. Veer of wind, commonly through about 4 points. Barometer steadies.	1. Veer of wind, commonly through 6 or more points, sometimes accompanied by line squalls. Barometer starts to rise.
2. Strong winds. Usually humid, stable air with much cloud and often intermittent rain or drizzle or snow. Moderate or poor visibility.	2. Short-lived heavy rain or snow, sometimes with thunder. Wind gusty and often temporarily increases.
	3. Cloud cover breaks. Usually unstable, comparatively dry air with cumulus-type cloud and scattered showers. Good visibility outside showers. Wind gradually eases and often slowly veers.

Approach of a Cold Front

No warning before actual appearance of large cumulus type cloud.

Passage of Cold Front

1. Veer of wind, commonly through 4 points or so, sometimes accompanied by line squalls. Barometer rises.

2 & 3. As with an occlusion.

Other Barometer Changes. The barometric changes set out in the above table, and described in the earlier parts of this chapter, occur due to frontal phenomena alone, and the assumption underlying them is that there are no other, simultaneously occurring, causes of barometric change. If there are other causes they will naturally affect the overall changes occurring. In practice there are three main additional causes usually operating. They are (1) deepening or filling up of the depression—i.e. in-

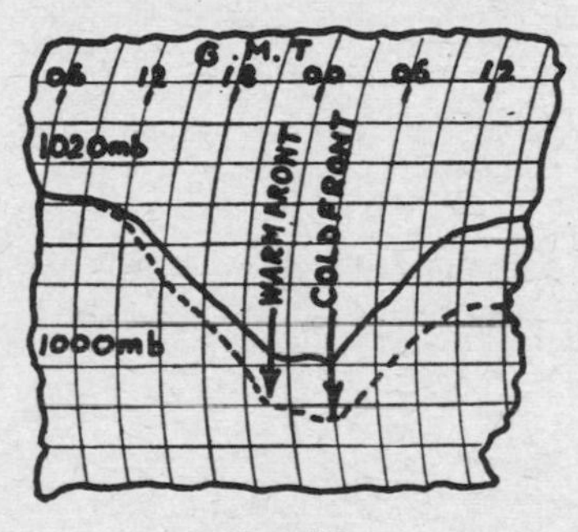

FIG. 28.

crease or decrease of the pressure at the centre of the depression; (2) movement of the centre of the depression in relation to the ship; (3) diurnal pressure variation.

Deepening or Filling Up of a Depression. So long as a depression is not fully occluded—i.e. so long as it has a warm sector—it usually deepens. Once it is fully occluded—i.e. once the cold front has caught up the warm front over its whole length—a depression usually starts to fill up. The effect of deepening is obviously to increase any existing rate of fall of the barometer; turn what would otherwise be a steady barometer into a slowly falling one; and decrease any existing rate of rise of the barometer. The full line in Fig. 28 shows typical barometric changes due to the passage of warm and cold fronts alone, while the broken-line curve shows what happens if, in addition, the depression is steadily deepening all the time. In Fig. 28 an exaggerated difference between the two curves is shown in order to make the nature of the difference plain. Note that the originally horizontal part of the full-line curve (corresponding to the warm sector) slopes down in the broken-line curve.

Movement in Relation to the Ship. Clearly so long as the path of movement of the centre of a depression and the track of the ship are such as to bring the centre nearer to the ship, this relative movement *alone* is enough to bring about a fall in the reading of the ship's barometer. Conversely, so long as the centre is moving away from the ship, this relative movement *alone* is enough to cause a rise in the barometer. In the case, common in British waters and the Northern hemisphere generally, where the warm sector extends more or less South of the centre of a depression, the centre of a depression will normally be getting nearer to a ship over which the fronts will pass until the ship is somewhere in the warm sector, and will recede thereafter. The main effect in such a case is, therefore, to increase the drop in the barometer taking place while the warm front is approaching and to increase the rise occurring after the cold front has passed.

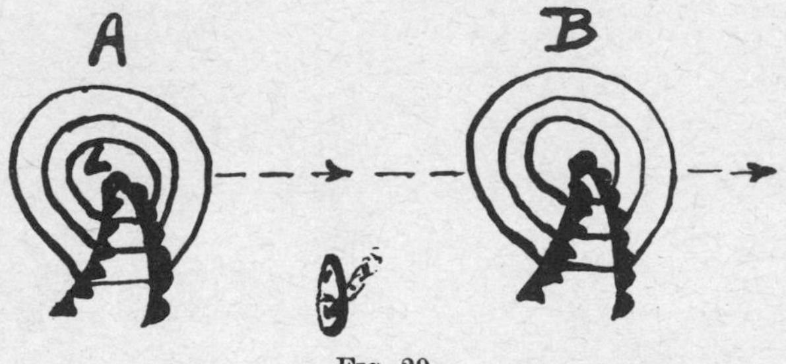

FIG. 29.

Fig. 29 illustrates this. In this figure the depression shown is assumed to move from A to B along an Eastward track passing the ship on the Northern side of her. It is assumed also (this would seldom occur in practice) that the depression does not alter in any way while this is happening. Clearly, while she is ahead of the warm front the ship will experience a fall of pressure due to the combined effects of (1) the substitution of warm air for cold overhead and (2) the fact that the centre of the depression is getting nearer. Once the centre has reached a point North of her, her pressure rise will be increased due to the fact that the depression starts to move away Eastward. Of course, ship's movement is part of the relative movement between depression and ship, but, in general, the speed of a little ship—six or seven knots, or something of that order—is too small relative to the speed of a lively depression to contribute much to relative movement. In the case of a fast ship, however, her own movements cannot be ignored, especially if she is steaming more or less directly towards or away from the depression centre.

Diurnal Variation. As the earth rotates, the atmosphere is subjected to an effect which is rather analogous to tide in the oceans. It results in a regular wave-like change in atmosphere pressure, with two maxima and two minima a day, quite apart from and additional to any changes due to meteorological causes properly so called. This regular variation is called diurnal variation of pressure. The maxima occur at about 10 o'clock and the minima at about 4 o'clock, morning and evening, local mean time, and the amount of variation depends on the latitude, being greatest at the equator (where it is 3 mbs.) and falling away towards the poles. At 60° it is only 0·4 mb. The three curves of Fig. 30 show diurnal variations for three different latitudes.

In the tropics diurnal variation can never be ignored because not only is it fairly large in itself but large in relation to the meteorological pressure changes experienced, which, in low

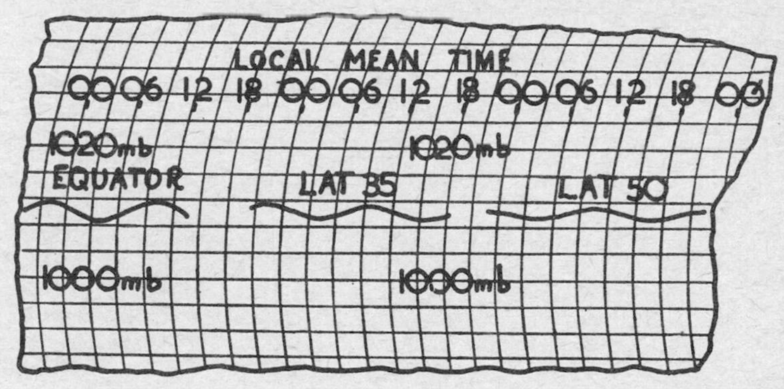

Fig. 30.

latitudes, need be only quite small to produce high winds. In low latitudes, if the barograph track noticeably departs from the diurnal variation curve it is a sure sign of a storm. In home latitudes, however—around 50°—and in higher latitudes, the total range of diurnal variation is under 1 mb., and this is small enough to be, in practice, ignored in relation to the much larger meteorological pressure changes which occur.

CHAPTER VIII

MOVEMENT OF DEPRESSIONS AND FRONTS: SECONDARIES

The factors governing the speeds and directions of movements of depressions and fronts are so complex and numerous that the master of the little ship is well advised to rely upon the directions and speeds given—as they often are—by the official forecaster in the shipping forecasts. It must be frankly confessed that the information ordinarily available in the little ship will be inadequate to enable more than the most general estimates of speeds and directions to be made aboard. Indeed, the great majority of the cases in which official forecasters go wrong are cases in which the errors are errors of estimation of speeds of movement. There are, however, a number of useful rules which can be given and which will enable the shipmaster who has a weather chart—and a later chapter will describe how he can draw a reasonable weather chart at sea from shipping forecasts—to make useful estimates of speeds and directions in those cases where the information is lacking from the shipping forecast.

Lows and Highs. Compared to lows (i.e. depressions), highs (i.e. centres of high pressure) are stable and move only slowly. A low whose centre is approaching the centre of a well-established high tends to slow up and/or to change direction as it approaches so as to go round the high in accordance with the anti-cyclonic air circulation—i.e. clockwise in the Northern hemisphere and anti-clockwise circulation in the Southern hemisphere. Thus a depression approaching the British Isles from the Westward and encountering a well-established high-pressure system over the Netherlands, Belgium and Central Europe is likely to turn North-East and affect only the Northern parts of the British Isles.

Fig. 31 illustrates this. The path of the centre L of the depression is likely to be somewhat as indicated by the arrow, so as to move clockwise round the edge of the high at H. In Northern latitudes in the neighbourhood of 45° to 60° any high with a

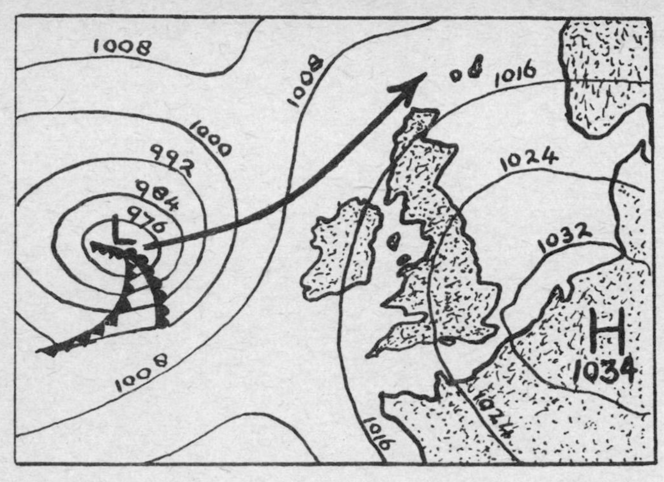

Fig. 31.

centre at a pressure of about 1025 mbs. or more may be regarded as established, and highs with centres at pressures much above this do not easily or quickly give way to lows.

Warm-Sector Depressions. So long as a depression has a good warm sector it usually deepens—i.e. the pressure at its centre gets lower—because it is being fed with plenty of warm air. In many cases depressions move more slowly while they are deepening, though the reasons for this are complex and rather obscure. Once they have become fully occluded they usually fill up (i.e. the central pressure rises and they die), and, again, it is a frequent experience that depressions speed up when they start to fill up. A reliable rule is that the direction of movement of the centre of a depression having a well-marked warm sector is parallel to the direction of the isobars in the warm sector.

This is illustrated by Fig. 32, which is a simplified representation of an actual depression. In this figure the frontal rain belts are indicated by shaded bands. The direction of movement of the centre L of the depression in this figure is shown by the thick black arrow.

Speeds of Movement of Fronts. The speeds of movement of different parts of fronts can be estimated with reasonable accuracy by measuring *along the front* the distance between the isobars—

note that the measurement must be taken along the front, *not* at right angles to the isobars—and setting off the separating distance along the geostrophic wind scale. Cold fronts and occlusions move at about the speed obtained from a geostrophic wind scale in this way. Warm fronts usually move at about 10 to 15 knots slower. As already explained, open-sea wind speeds are about two-thirds of geostrophic wind speeds, and since the master

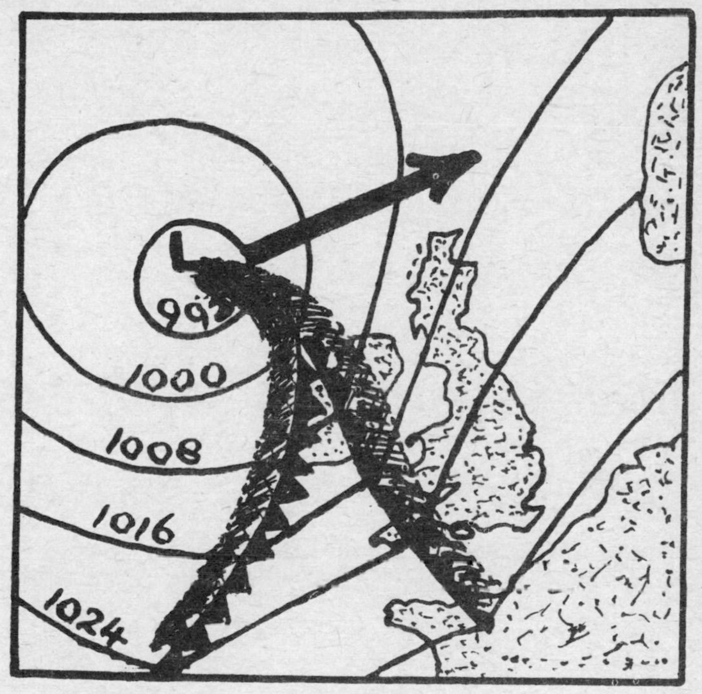

Fig. 32.

of the little ship is not really interested in geostrophic wind speeds (which are speeds at 1500 ft.) it is considered more useful to provide him with a chart having a modified geostrophic wind scale giving open-sea wind speeds direct. Chapter XI, entitled "Making the Chart at Sea", includes a recommended form of chart with a modified scale of this nature. When using such a scale, speeds of cold fronts and occlusions may be estimated by measuring the

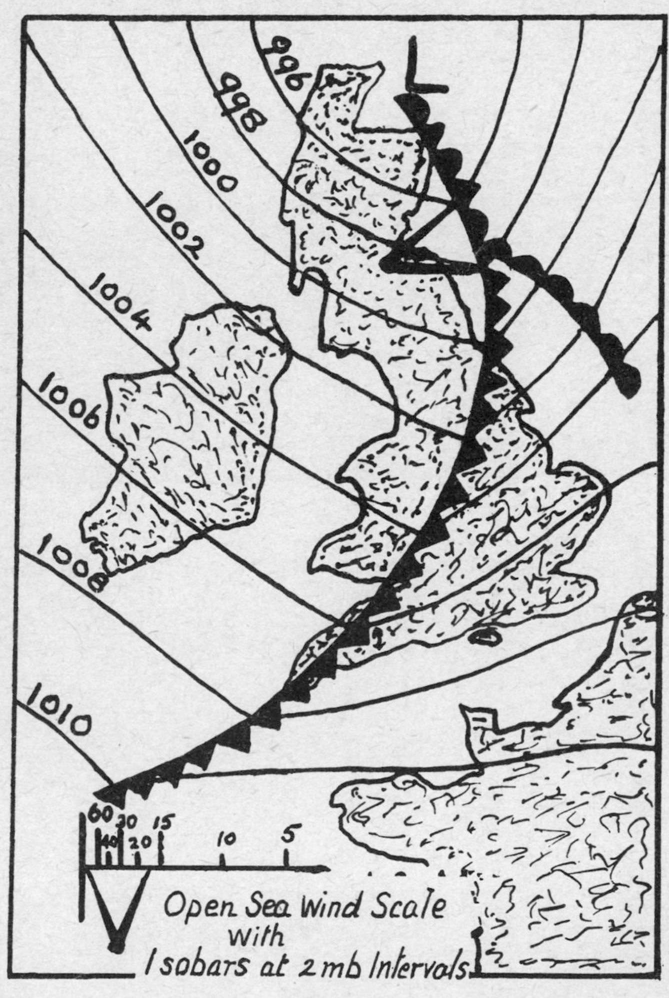
996
998
1000
1002
1004
1006
1008
1010
L
60
30
15
40
20
10
5
Open Sea Wind Scale
with
Isobars at 2 mb Intervals

FIG. 33.

isobaric spacing along the front, setting it off on the scale and adding 50% to the scale reading, warm front speeds being, of course, again 10 to 15 knots, or thereabouts, slower.

In Fig. 33 the portion of the front between the points of the dividers may be estimated as moving parallel to itself at a speed of about 25 knots—i.e. 50% more than the reading as given by the dividers on the open-sea wind scale.

Trailing Fronts. In general, of course, the isobars near the centre of a depression are nearer together than those farther out. Since frontal speeds, like wind speeds, depend on isobaric spacing, different parts of the same front move at different speeds, parts near the depression centre moving faster than those farther out. Accordingly fronts tend to trail, the trailing parts cutting the isobars more and more obliquely. In the case of a cold front or occlusion very considerable trailing up to the point where it has a part which is parallel to the isobars is very common indeed. The part of a cold front or occlusion which is parallel to the isobars is stationary and the end of it commonly continues into the end of a warm front belonging to the next depression.

Families of Depressions. Polar Front. Ridge of High Pressure. Single, isolated depressions in temperate latitudes are the exception rather than the rule. In such latitudes depressions usually come in sequences or "families"—commonly at least three in succession, sometimes as many as six in the case of depressions reaching the British Isles from the North Atlantic. A convenient and, it is believed, accurate way of looking at what happens is to regard the depressions as forming along a general boundary line between cold North or North-West air originating in polar regions (the Northern hemisphere is assumed here) and warm, humid, approximately S.W. air from tropical or sub-tropical regions. This general boundary is commonly, if somewhat loosely, called the "*Polar Front*". It extends approximately North-East across the Atlantic and is farther North in summer than in winter. "Waves" form on it, each "wave" having a leading edge which is the warm front of a depression and a trailing edge which is a cold front (or occlusion) which trails into the warm front leading edge of the next depression.

Fig. 34 illustrates a typical North Atlantic family of depressions. Each depression L in the family tends to follow a path a little to the South of the depression in front of it, and each depression usually gradually loses its warm sector and becomes occluded as it gets farther and farther Eastward. The way in which the cold front of each depression trails into the warm front of the one behind it is clearly shown. Between each successive pair of depressions is a *ridge of high pressure* marked R. Speeds of movement of such a family, of course, vary widely, but a typical figure for the time taken for a complete family to pass Eastwards or North-Eastwards over a given place in the path is 5 or 6 days. A few wind directions are shown in Fig. 34 and particular attention is drawn to the way in which these directions change as they follow round the ridges marked R. If a ship

FIG. 34.

which has had a depression move past her finds the N.W. wind behind the cold front back to West or South of West, and her barometer cease to rise and start to fall again, she can be pretty confident that another depression in the family is on its way towards her.

Secondary Depressions. A secondary depression is a small centre of low pressure formed within a larger low-pressure system. "Secondaries", as they are usually called, may vary in intensity from one which appears on the chart as a mere opening or curving of the isobars, to one which is a complete low-pressure system with closed isobars round a well-marked central low and its own air circulation round it. It may even have its own well-marked frontal system; though, commonly, the size of a secondary is, in its early stages, too small for fronts to be clearly detected and delineated even if they are present, as they probably are.

Although much smaller, at any rate initially, than the primaries with which they are associated, secondaries can be both violent and fast-moving and some of the highest winds which have been experienced in temperate latitudes have been due to secondaries.

The great majority of secondaries form on the equatorial side of primary depressions—i.e. in the Northern Hemisphere they form somewhere in the Southern half of a primary depression. An opening of the isobars usually constitutes the first appearance of a secondary on the chart, and if such an opening is seen on any chart it is wise to keep a close watch on the weather, taking

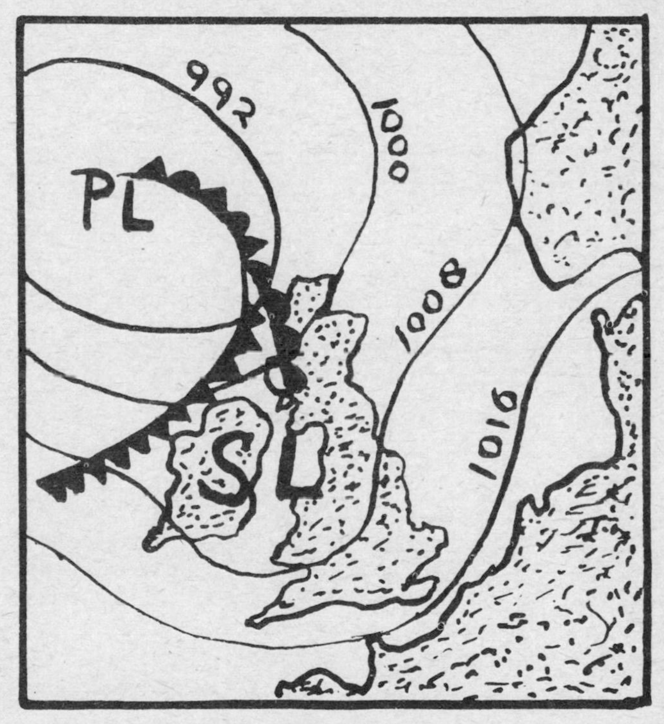

Fig. 35.

particular note of any unexpected fall in the barometer, and drawing charts (as will be explained later) from each shipping forecast as it comes in, for a secondary can develop very rapidly, advance at a very high speed and produce most dangerous, if short-lived, gales.

Fig. 35 shows at SL a secondary which is not as yet of much intensity and which has formed at SL, where the 1008 mb. isobar opens away from the 1000 mb. isobar in the circulation of a primary whose centre is at PL. Such a secondary, even if not in

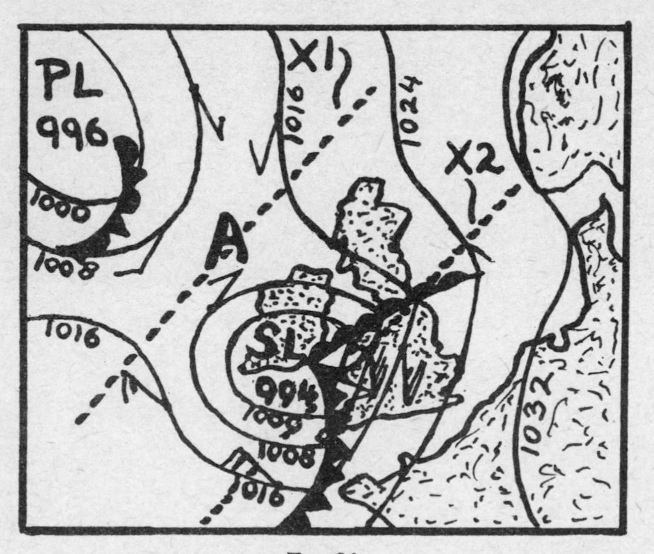

FIG. 36.

itself very active, should never be ignored if it appears on the chart, because it can easily and quickly develop into something of the nature of that shown in Fig. 36. Here there is shown a deep secondary SL—actually deeper than the primary PL—which has become the dominant centre on the chart, and is deepening rapidly because of its well-developed warm sector.

Movement and Development of Secondaries. Secondaries move in the general air circulation of primaries as well as sharing in their movements. Thus, in Fig. 36 the secondary at SL will swing anti-clockwise round the main centre PL as that centre moves along—about North-East in this particular example shown. Usually the primary slows up somewhat during the development and deepening of an associated secondary. It is uncommon for a secondary to swing through more than about half of one complete rotation round the primary. What usually happens with a deepening secondary is that, as it deepens, the primary fills up, and by the time the secondary has swung through about a quarter or a third of a circle round the primary the secondary has become the main centre and what was originally the primary is disappearing. This is the position pictured in Fig. 36.

Secondary-Depression Winds. It is very important to a ship on the equatorial side of the main centre of a depression with a secondary in its circulation to know where she is with relation to the path of the centre of the secondary, because the winds she will experience will be totally different depending upon whether she is North or South of the secondary centre. Look again at Fig. 36. It will be observed that there is an area A of light variable winds between the two centres PL and SL, with roughly South-Westerly winds on the N.W. side of that area and roughly North-Easterly winds on the S.E. side of it. If the line X1 represents the path along which the whole system passes a ship, she will experience roughly Southerly winds followed by a period of light variable winds, followed by Westerlies or North-Westerlies. If, however, the line X2 is the relative path, the South-Easterly winds will be followed by strong Southerlies, in turn followed by strong Westerlies. It is important to notice that the winds experienced as the centre of the secondary passes by on the North-West side of the ship will be much stronger than the winds which would have been experienced at that distance from the primary centre were there no secondary, for the presence of the secondary not only reduces the strength of the wind in the area A *but correspondingly increases the wind strength on the equatorial side of the secondary centre.* Violent if short-lived gales close on the equatorial side of centres of secondaries are by no means uncommon, and gales or near gales are often caused by secondaries in parts of low-pressure systems where, were there no secondary, moderate or even light winds would be expected.

Secondaries can form and develop with great rapidity and commonly move very fast, because the movement of a secondary is a compound one made up of the movement of the associated primary *plus* its own movement as it swings forward round the primary centre; and although the primary usually slows quite a lot while the secondary is swinging round it, the compound movement is, nevertheless, a pretty fast one. Because of this liability to rapid development and high speed, secondary depressions commonly give much less warning on the barometer than do ordinary depressions. It is this characteristic that underlies the old couplet:

> "*Long foretold, long last;*
> *Short notice, soon past.*"

CHAPTER IX

AIR MASSES

Air-Mass Weather. We have seen that weather associated with fronts is due mainly to the fact that, at a front, one air mass is replacing another. Frontal weather may therefore fairly be regarded as "change of air mass" weather. In the absence of fronts, or between fronts, the weather experienced is determined primarily by the physical properties of the air mass which is present. Such weather may be termed "air-mass" weather.

Polar and Tropical Air Masses. The most convenient way of classifying air masses is in terms of their origin, and for practical purposes it is sufficient to divide air masses into two main classes, one cold and the other warm. The main cold air mass is called *Polar Air*, which is air originating in polar or near polar regions. The main warm air mass is called *Tropical Air*, which is air originating in tropical or equatorial regions. Many meteorologists distinguish between *Polar Air* and *Arctic Air*, meaning by the former air which comes from centres of high pressure in high latitudes towards the poles, and reserving the latter term for air coming actually from the great icefields at the poles. The differences between Polar Air (with this meaning) and Arctic Air are not, however, very great, and for the purposes of practical weathor work in little ships it is entirely adequate to drop the distinction and use the term Polar Air to cover both. Similarly, many meteorologists reserve the term Tropical Air for air which comes from regions in and near the tropics rather than from equatorial regions and describe air from the latter regions as *Equatorial Air*. Again, however, the differences between Tropical Air (with this meaning) and Equatorial Air are not, for our purposes, enough to bother about and the term Tropical Air will be used herein for both.

Modification of Air Masses. Maritime and Continental. It is the great characteristic of air masses that they modify their physical properties—primarily temperature and humidity—in their lower layers comparatively slowly as they move from their areas of origin. The extent to which such modification takes place obviously depends largely on the nature of the earth's surface over which the air mass moves: obviously the changes brought about in the lower layers of a polar air mass (for example) after it has blown over hundreds of miles of sea will be different from those occasioned by blowing for the same distance, and in otherwise similar conditions, over land. This leads to the only sub-classification of air masses with which we need trouble ourselves, namely *maritime* (usually designated by the preliminary letter *m*) for an air mass which has reached us over a predominantly ocean path and *continental* (usually designated by the preliminary letter *c*) for an air mass which has reached us over a predominantly land path. We may therefore distinguish between four main sorts of air mass: two cold (i.e. colder, on the average, than the surfaces over which they move) and two warm (i.e. warmer, on the average, than the surfaces over which they move). The two cold ones are Maritime Polar (mP) and Continental Polar (cP) and the two warm ones are Maritime Tropical (mT) and Continental Tropical (cT).

General Nature of Cold Air Masses. The two polar air masses are, in their original state, stable (especially in their lower levels), of low temperature, and contain comparatively little moisture. Indeed they commonly exhibit temperature inversions, a "temperature inversion" being the state of affairs in which the temperature at a particular height is actually higher than that at a lower height. As such an air mass advances into warmer latitudes it is heated from below. A larger and larger temperature lapse rate therefore develops in the lower layers because the air at some distance above the surface remains cold while the air at the bottom becomes warmer. The lower layers therefore tend increasingly to become unstable and this instability extends gradually upwards. Any temperature inversions originally present are destroyed. As a result of the instability thus developed, smoke and dirt generally disperse and visibility is good outside precipitation. By the time mP or cP air has reached British or Western European waters it is usually markedly unstable. In the case of cP air—i.e. polar air which has reached us by an overland route—very little moisture has been picked up en route and the relative humidity is low so that cloud amounts are small, what little cloud cover there is tends to be very broken, of cumulus type and at a good height, and there is nothing much in the way of precipitation. The principal difference between cP and mP air is that the latter picks up considerable quantities of moisture in its lower levels in the passage over a long ocean path. Again, however, there is heating from below as the mass advances to warmer latitudes, so that instability develops in the lower layers. This causes the moisture picked up to be taken higher and higher into the mass. Cumulus-type clouds accordingly form and grow into large cumulus or thunder-clouds and precipitation of the shower type is common. The cloud base is usually fairly high and visibility outside showers is good. Another difference between cP and mP air is that, because sea temperature hardly changes at all between night and day, diurnal variation of the cloud characteristics of mP air do not amount to very much, whereas, with cP air, there is often considerable diurnal variation of cloud because land rapidly heats up when the sun is on it by day and rapidly cools off during the night. Summarising in tabular form the principal features possessed by mP and cP air masses by the time they have travelled a long way from their origins into temperate latitudes, we have:

mP Air	*cP Air*
High lapse rate with considerable instability.	High lapse rate with considerable instability.
Gusty winds if at all strong (due to instability and squally showers).	Some gustiness of the stronger winds.
Large cumulus-type clouds with breaks.	Scattered cumulus-type clouds not often very large (because not much moisture).
Precipitation of the shower type.	Not much precipitation, but what there is is of the shower type.
Fairly high cloud base (seldom below 1500 ft.)	High cloud base (usually above 2500 ft.)
Not much diurnal variation of cloud or showers.	Marked diurnal variation of cloud and showers with maxima in the afternoon (because then the land will be at its hottest and produce
Very good visibility outside showers.	

	instability to the highest levels).
	Generally good visibility, but haze sometimes occurs.

General Nature of Warm Air Masses. There are two main sources of warm air reaching British and Western European waters. The commonest source is from the area of high pressure near the Azores. This air, which travels over an ocean path and is therefore a maritime air mass (maritime-tropical: mT), is originally warm, fairly stable and of high moisture content in the lower layers. As it advances into more temperate latitudes it cools from the bottom and becomes increasingly stable and of lower lapse rate. The high stability stops the formation of up-currents and confines cooling to the bottom layers. Accordingly the production of a temperature inversion at a comparatively small height is quite common. If the bottom layers become sufficiently cool to bring down the air temperature near the surface to below the dew point, fog is formed, and much fog is caused in this way in home waters. The cloud is mostly stratiform, often low, and commonly in an unbroken sheet or in a sheet with only a few breaks, and precipitation tends to be of the drizzle type.

The second main source of warm air reaching British and Western European waters travels over a predominantly land path and is therefore continental—cT (continental tropical). It is essentially, so far as British waters are concerned, a summer air mass and is hardly ever encountered in those waters in winter. It is warm and dry and moderately stable. Because of the low moisture content there is very little cloud. Visibility is usually moderate to good. Tabulating the main properties of the two principal warm air masses mT and cT, we have:

mT Air	*cT Air*
Stable, with low lapse rate or temperature inversion in the lower layers.	Moderate stability. Not much wind turbulence. Very little cloud or precipitation.
Winds steady and without turbulence unless strong enough to produce turbulence mec-	Moderate to good visibility.

hanically from high waves or land obstructions.
Stratiform cloud.
Drizzle-type precipitation.
Low cloud base, sometimes right down to surface.

British Waters Air Masses. By far the commonest air masses in British waters are the two maritime air masses mP and mT. In these waters the former is nearly always the air mass to come in behind a cold front or an occlusion and has a wind direction

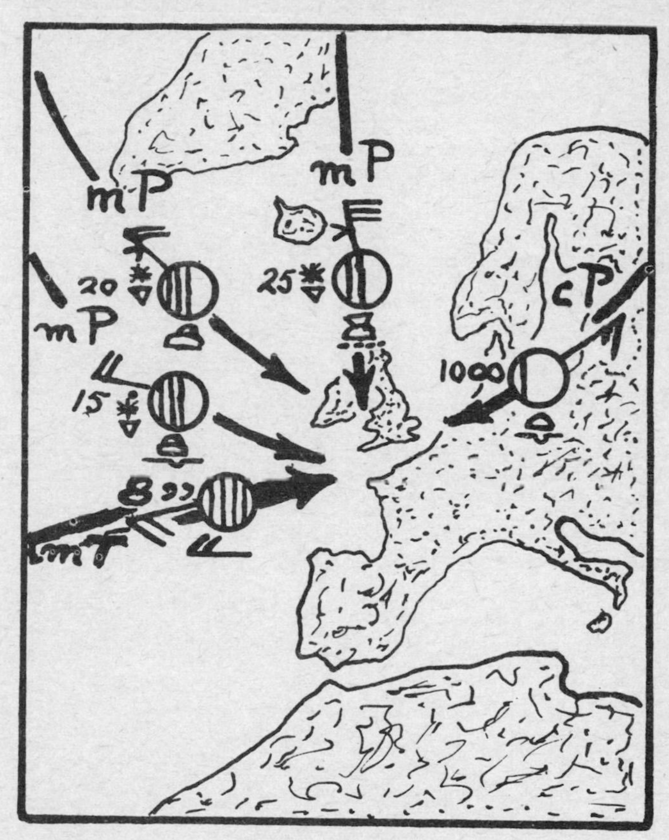

FIG. 37. WINTER.

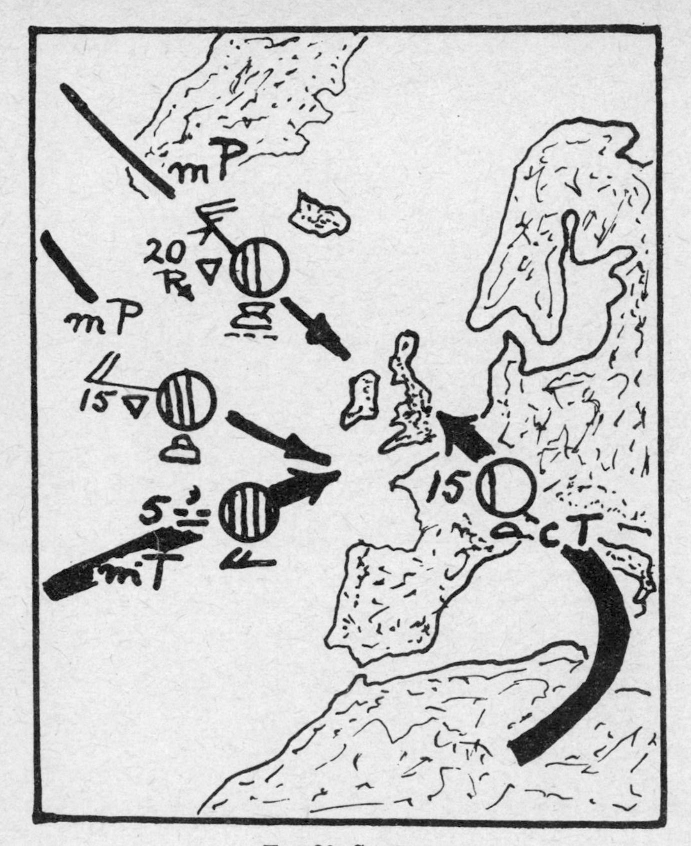

FIG. 38. SUMMER.

usually between W. and N.—commonly N.W. mT air is nearly always the air mass in a warm sector behind a warm front and commonly has a wind direction between about S. and W.—most usually S.W.

Of the two "continental" air masses, cP and cT, the former is mainly a winter air mass, coming in a little North of East as a cold, dry mass with little, if any, precipitation and moderate visibility, not infrequently with haze, while the latter is almost exclusively a summer air mass, coming in somewhat East of South, hot and dry and usually in a so-called "heat wave".

Obviously the properties of air masses will differ in summer from those possessed in winter and also there will be small differences between generally similar air masses reaching us over slightly different paths. Fig. 37 shows the main British waters air masses in winter with their general weather characteristic represented in the conventional manner described in Chapter III. Fig. 38 is a similar representation of the principal British summer air masses with their general weather characteristics. It should be stressed that the weather characteristics shown in these two figures are only typical, average characteristics: there are, of course, considerable variations.

CHAPTER X

OTHER ISOBARIC PATTERNS

In preceding chapters considerable space has been devoted to the formation and behaviour of depressions and their associated fronts because depressions play a very large part in determining weather in temperate latitudes and, in a large number of weather charts, constitute the most important chart features. However, there are, of course, other forms of isobaric patterns to be recognised and considered.

Anti-cyclone. Anti-cyclones, or "highs", are, as the name implies, areas of high atmospheric pressure enclosed by isobars. They are commonly the most stable things on the chart and, in temperate latitudes, may remain more or less in the same place for days or even weeks. They are often irregular in shape and usually very extensive and, as already stated, they tend to "remain put" and cause depressions to slow up or deflect and go round them—commonly both. They are essentially quiet in their weather conditions, and strong winds in the central parts of anti-cyclones are extremely rare. Where, however, a succession of depressions passes round part of the periphery of an anti-cyclone, steep pressure gradients may be set up and high winds and gales, often long-lasting, can occur, though they are not very common. Such gales are sometimes referred to as "anti-cyclonic", though the depressions have at least as much to do with them as the highs themselves.

Fig. 39 illustrates this sort of thing. There are certain parts of the world where there are more or less permanent anti-cyclones which "dither about", expand and contract, strengthen or weaken, but are more or less always to be found somewhere in the same general area. One of these is the so-called "Azores High", an extensive area of high pressure in the Southern and Eastern part of the North Atlantic. In Fig. 39 it is marked with the reference "Az.H". A series of deep depressions each marked "L" is shown moving along the Northern edge of this anti-cyclone.

It will be noted that the isobars are quite close together along the Northern edge of the anti-cyclone where the wind arrows are shown, giving persistent strong winds which, though they may ease a little temporarily after one depression has passed and before the next one replaces it, will remain strong until the whole system of depressions has gone by. In the Southern hemisphere the more or less permanent strong winds and gales of the "roaring forties" occur along the edges of the quasi-permanent high-pressure belts in latitudes around 40° or 45° or so South.

Fig. 39.

Subsidence. As has already been seen, the air circulation round a high is clockwise in the Northern hemisphere and anti-clockwise in the Southern, with, in both cases, a small component (about 1 compass point) of wind direction outwards. Because of this outward component of wind direction an anti-cyclone is always losing air at the lower levels and this air is replaced by air which enters at very high levels and descends in the middle parts of the system. There is thus a downward flow of air—or *subsidence* as it is called—in the central parts of an anti-cyclone. The air as it comes down is warmed as a result of compression (as already explained) and this, of course, operates to resist the formation of cumulus cloud or precipitation. Rain is rare in a "high" and if precipitation does occur it is nearly always drizzle. Vertically developed clouds of any great height are also rare, though cloud, mostly of the stratus and the strato-cumulus type, is not uncommon, especially in winter, giving overcast or nearly overcast skies. In fact, in an anti-cyclone there tends to be either a lot of cloud or very little, and, roughly speaking, anti-cyclones can be classified as "fine" or "cloudy". In summer, in British and near European waters, anti-cyclones usually give fine weather—a North-Westerly extension of the Azores High is often the dominant feature of the British weather chart for fine warm summer days—though it is a mistake to assume that fine weather is a necessary accompaniment of an anti-cyclone even in summer.

Visibility is usually only moderate or hazy because the subsidence resists the formation of up-currents to clear the air. If the anti-cyclone is of the "fine" (i.e. cloudless or nearly cloudless) type and the winds are light or calm, fog and ground mist are apt to occur over land at night because, with a cloudless sky, the land cools rapidly at night and may bring the bottom levels of the air in contact with it below the dew point. Such fogs are usually not of any great thickness and therefore dissipate rapidly during the forenoon as the sun heats up the land again. They seldom drift any great distance to sea, though they may hide harbours, shore lights and coastlines. In winter, however, night fog formed as above described may become thick enough not to disperse during the following day, and if this happens the day temperature on land will remain cold due to the fog "blanket" on it, and persistent fog, which can reduce visibility in estuaries to nearly zero, will occur—though, again, this sort of fog seldom drifts far to sea.

Ridge of High Pressure. A *ridge* of high pressure is a tongue or wedge-like extension of an anti-cyclone extending up between two lows. The older term for it—still occasionally used—is a "*wedge*".

Fig. 40 shows "ridges" at R1 and R2, between lows at L and L and extending towards one another from highs at H and H to the North-East and South-West of the pictured area. In British and adjacent waters the commonest kind of ridge is one which, like the ridge R2, projects more or less northerly out of an anti-

FIG. 40.

cyclone. As will be seen, as such a ridge passes, the wind, which has previously veered to, say, between West and North, eases off and backs to West and then to South-West or South. The barometer, which has been rising, steadies about the time of the beginning of the backing of wind and then falls. This combination of a backing wind and a steadying and then falling glass is, in itself, noteworthy indication of another depression on the way and, if accompanied by increasing cirrus, is a fairly certain promise of one.

The movement of a ridge takes place with and is governed by the movements of the lows on the two sides of it. "Ridge weather" is probably the best weather in the book—almost always fine, like anti-cyclonic weather but without the persistent cloud sheets commonly accompanying winter "highs"—but, alas, the fine weather does not last long, for the ridge soon passes. The usual weather sequence in Northern temperate latitudes is fresh N.W. winds (or thereabouts) with showers and bright-blue sky intervals; showers die away; wind eases and starts to back; cirrus comes in fast from between N.W. and N.; wind falls light and high cloud comes in from the westward; wind backs further and slowly increases . . . and then the sequence of weather due to the new depression behind the ridge.

Trough of Low Pressure. This term, which is still (rather regrettably, I think) fairly often used in broadcast forecasts is, practically speaking, only another general term for a front. A *trough of*

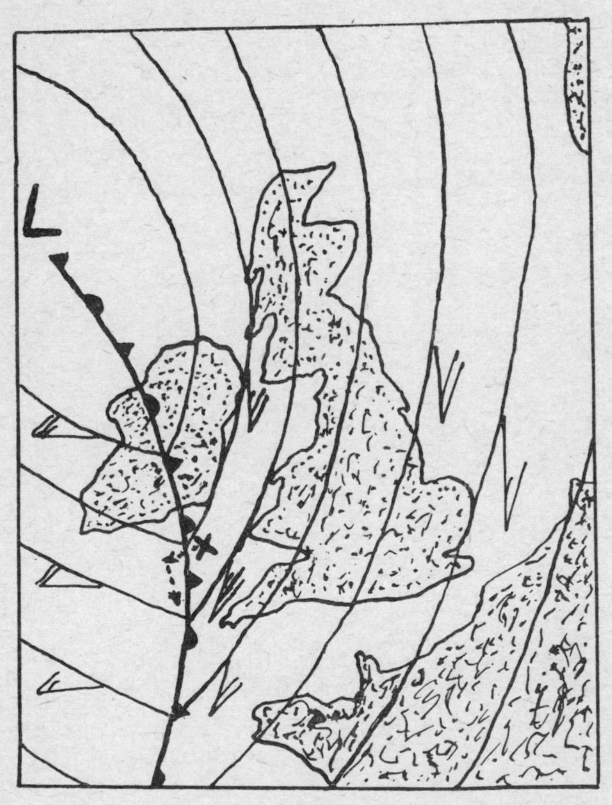

Fig. 41.

low pressure may be a warm front, a cold front, or an occlusion. In present broadcast forecast practice in some countries—the United Kingdom is one—there appears to be something of a tendency to use this term to describe what used to be known by the now obsolete but descriptive term of "*V-shaped depression*".

V-shaped Depression. A *V-shaped depression* is an occluded depression in which the change of direction of the isobars at the occlusion is so great and sharp that the isobars here take up a V-shape—hence the name.

Fig. 41 shows a typical V-shaped depression. The master of the little ship under sail should take particular note of the wide angles between the wind directions ahead of and behind the front on the one hand, and the direction of extension of the front itself on the other. Because of this, a ship at, say, X, in Fig. 41, desiring to head South to round Land's End, the South-West extremity of England, and go East up the English Channel would find it pay her handsomely not to yield to the temptation to head full and by as near South as she will lay, for if she does she will not only be close-hauled a long time but will also be a long time in the bad weather of the front, since she will cross it obliquely. If, instead, she sails free and heads a shade South of West to cross the front more or less at right angles, she will close it and cross it rapidly and, by shaping up as indicated by the broken-line arrows and turning South (or thereabouts) as soon as she finds the Westerly winds behind the front, will soon get fair winds and better weather. In the case of a slow-moving front, much can be gained in this sort of way.

Col. The *col*, though recognised as a typical isobaric pattern, is really only a gap or space between other patterns, being, in fact, the space between two highs and two lows which, together, surround it. Going round a col, we have "high", "low", "high", "low" in turn, and providing the highs and lows alternate round the col they can be in any positions with relation thereto—e.g. high pressure to North and South and low pressure to East and West (probably the commonest case in British waters), or high pressure to East and West and low pressure to North and South. The col itself is therefore the meeting-place of the different air masses belonging to the systems which surround it and the weather experienced will depend on those air masses and, as will be obvious, can easily be different in different parts of the col.

Fig. 42.

Fig. 42 shows a typical col, indicated by the letter C, as frequently experienced in British waters, for example. It is virtually impossible to generalise about the weather in a col—since this weather depends on the air masses which border it and is also likely to be different along different borders—beyond the general statement that it is an area of light winds or calms in its central parts. For this reason if the moisture content and temperature gradient of one (or more) of the bordering air masses are such that fog can develop, as is often the case in winter, fog will occur in the col, and the reputation which winter cols have of being foggy is not undeserved, though the facts do not justify the general statement often heard, that the presence of a col in winter is enough to warrant an expectation of fog. In summer, over land, the light winds and calms are favourable to intense solar heating of land, and therefore, if one or more of the bordering air masses is moist enough and the heating is great enough, thunderstorms will form and may drift out. In fact thunderstorms fairly frequently occur in summer cols over land in the afternoon or evening, but, as with the case of fog, a general statement that a summer col means thunderstorms is not justified. The fact is that weather

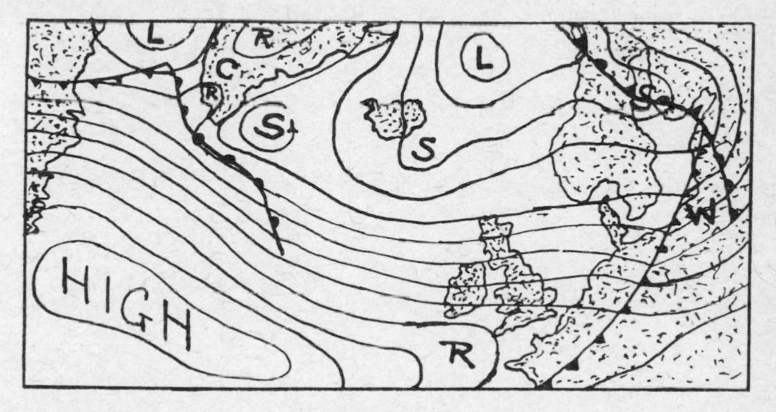

Fig. 43.

in a col is governed by the surrounding air masses, and col weather forecasting is really only air-mass weather forecasting.

The movement of a col depends upon and is governed by the movements of the highs and lows which encompass it, and, from its very nature as a mere space between other systems, it is essentially a transitory and impermanent thing. Very few cols retain their form or even their identifiable existence for any great length of time.

Fig. 43 is a simplified actual North Atlantic weather chart (the time of the year was late autumn) which is added as a practical example showing a good many isobar forms. There is a low L to the N.E. of Iceland, with three secondaries indicated by the letters S. It will be noted that these secondaries are shown, as is common with very large area charts on which, necessarily, only a few of the isobars appear, merely as "loops" or "pockets" in the isobars. There is, of course, the usual Northern hemisphere anti-clockwise circulation round each secondary, as is indicated for one of them (the one to the S.E. of Greenland) by a curved arrow. It will be seen that the secondary over Finland is partly occluded, with a warm sector W over Poland and E. Germany. The letters R denote ridges or wedges of high pressure and C is a col.

CHAPTER XI

MAKING THE CHART AT SEA

We have now reached the stage at which we can claim to have a general working knowledge of the mechanism of the weather, the nature and behaviour of the principal isobaric patterns to be found on the weather chart, the way in which those patterns develop, change and move, and the nature and behaviour of fronts and air masses. All this, however, is of little practical use to the master of the little ship at sea unless he has a weather chart to which this knowledge can be applied, for without such a chart his information is limited to what his barometer tells him and what he can see in the comparatively insignificantly small area within range of his eyesight. Now, of course, no ordinary little ship has, or can be expected to have, the communication equipment to enable it to obtain, at sea, the detailed charts which the official forecaster ashore has; indeed, even warships with their elaborate apparatus and large staffing of officer specialists cannot obtain such charts. Nevertheless, it is entirely practical, and, with a little experience, quick and easy, for the little ship within broadcast receiver reach of official broadcast shipping forecasts, with no communication equipment beyond an ordinary broadcast receiver, to use such shipping forecasts to draw simplified weather charts which, in main outlines, are astonishingly like those which the professional forecaster ashore has prepared for his own use. These simplified charts will be found really useful and, indeed, in an overwhelming percentage of cases, entirely adequate for practical purposes. I have experimentally used the method now to be described over the past twenty-five years in conjunction with the shipping forecasts of the British Meteorological Office as broadcast by the British Broadcasting Corporation and have yet to find a case in which the chart obtained thereby is seriously wrong, misleading or inadequate. In the remainder of this chapter the method is described in detail as used with these particular forecasts, but the principles of the method are general and it can be used, in a manner which will be obvious, with similar broadcast forecasts serving other parts of the world.

B.B.B. Shipping Forecast. At the present time, shipping forecasts, originating with the British Meteorological Office, are broadcast four times daily by BBC Radio 2 (long wave only: 1500 metres = 200 Kc/s) and can be used, as will be described in this chapter, for drawing weather charts at sea. The broadcast times, which have been changed in the past and may be again, are, at the time of writing (British Summer Times):—

02 02	every day.
06 30	every day.
11 55	Sundays only.
13 55	Monday to Saturday inclusive.
17 55	every day.

Make-up of Broadcast. Each broadcast starts with gale warnings, if any (gale warnings are also broadcast as soon as possible after issue and are repeated at the immediately following hour); follows this with a "general synopsis" which is a word picture of the weather chart giving positions and other details of pressure systems and major fronts, usually including the central pressures of pressure systems and often including the expected directions and speeds of movement of pressure systems and fronts or their expected positions at stated later times; follows this with forecasts, for the next 24 hours, of wind direction and force, weather, and visibility for each of the different sea forecast areas; and concludes with the latest reports from coastal and sea observation stations in the following list:—
Wick (W), Bell Rock (BR), Dowsing (D), Galloper (G), Varne (VN), Royal Sovereign (RS), Portland (P), Scilly (S), Valentia (V), Ronaldsway (R), Prestwick (PW), and Tiree (T). (The bracketed letters are those used to identify these stations in the chart form of Fig. 44). The station reports include wind direction and force, weather, and visibility and, in nearly all cases, also the barometer reading in millibars

and the barometric "tendency" (i.e. what the barometer is doing —e.g. "rising" or "falling").

General Method. My general method of using these broadcasts to draw the charts is as follows. I have produced a blank chart form which shows the sea forecast areas and the observation stations in the above list, and is also marked with a scale of miles, a modified geostrophic wind scale giving open-sea wind speeds directly for isobar spacings of 2 mbs., and a conversion scale from inches to millibars (in case the ship's barometer is marked in inches). Two of these blank forms are kept permanently, back to back, in a transparent, weatherproof Perspex or similar flat envelope which will take china-glass pencil markings which can be subsequently wiped off. The blank chart forms can be any convenient size. I find in practice that a form measuring approximately $12\frac{3}{4}'' \times 10''$ is entirely satisfactory. It goes nicely into one of those flat Perspex map cases of the sort commonly used by the military. As the broadcast comes in, the information is marked, in china-glass pencil, on the flat Perspex case in the correct positions over the appropriate sea areas which can be seen through it, using the logging symbols described in Chapters II and III. If, as commonly happens, a shift of wind is forecast for any sea area, the new wind is shown by a dotted wind arrow. Thus a forecast for "Humber" as follows—"wind South-West Force 4, becoming North-West Force 5; showers; good visibility outside showers"—would be shown in the middle of the "Humber" area, thus:

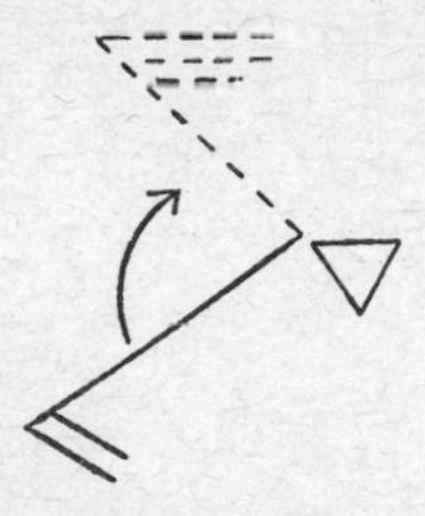

I do not consider it worth while marking up visibility for a sea area if it is only given as good or moderate, but I mark haze, mist or fog by the appropriate symbol, or, in the few cases in which visibility is given in actual distance (miles), I mark in the figure given. In addition the letter "G" is put into any area mentioned, in the beginning of the forecast, as having a gale warning. I show fronts as mere lines: it will soon be found that this is good enough, and after a little practice it can be seen from its position on the chart whether a front is warm, cold, or occluded, without marking its nature. However, to begin with at any rate, it is better to show the nature of the fronts, and this can be done by drawing a zig-zag line for a cold front, a rounded bump line for a warm front, and a plain line for an occlusion, thus:

Cold Front *Warm Front* *Occlusion*

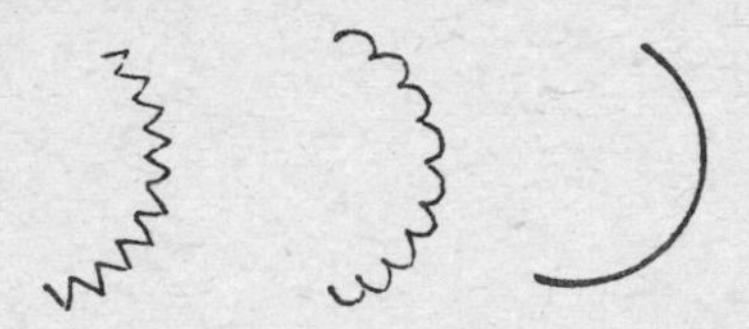

Information with regard to the observation stations is also plotted by means of the symbols given in Chapters II and III, making a special point of recording the barometer readings and tendencies ("rising", "falling", etc.). With very little practice it will be found quite easy, without hurrying yourself at all, to get all the information down on the chart as it comes in. When this has been done, the wind, weather and barometer observations of your own ship are added in the appropriate position to give you one more reporting station on the chart. For this to be done correctly your own barometer must, of course, be kept corrected as already described in Chapter II.

By the time the broadcast ends you will have in front of you a chart marked with forecast wind directions and speeds and weather for the sea areas; the positions, values, and expected movements of pressure systems; positions and expected movements of fronts; and a number of station observations (including that of your own ship) with their barometer readings and tendencies.

It is often important to remember that the general synopsis is not for the same time as the barometer readings. Thus, in the 1355 broadcast the general synopsis will be for midnight and the station observations (usually) for 1200. The times are stated in the forecast. Therefore, if the weather is moving rapidly—as in the case of a deep, fast-moving depression—the first thing to do is to shift the main pressure centre (or centres; but there is usually only one) by the movement it has had in the interval between the time of the general synopsis and the time of the station observations, so that the synopsis and the station observations will "fit". In many cases, however, the weather will not be moving rapidly enough to make this necessary.

The last step in the process is to draw the isobars, making them "fit" the plotted observations of pressure and the plotted positions of centres and fronts, using, if necessary, the modified geostrophic wind scale to ensure that the isobaric spacing "fits" the wind speeds given. Nineteen times out of twenty this is (with a little experience) consummately easy to do, for, although there are not many station pressures given, there are enough, taken in conjunction with the general synopsis, to give you your chart quickly and easily. I generally reckon to have my chart finished less than five minutes after the end of a broadcast.

When the next five-minute broadcast comes in, the case, with the chart forms in it, is turned over, and the new chart made on the back, and then, just before the succeeding broadcast is due, the china-glass markings giving the first chart are wiped off with a rag so as to leave the first side of the case clean and ready for the third chart. In this way, continuity is maintained and you always have two successive charts on record except when you are actually drawing one.

Chart Form. Fig. 44 is a reproduction, to scale, of my own chart form for use in British waters. The modified geostrophic wind scale and the sea-mile scale are in the bottom right-hand corner and the millibar-inch conversion scale on the upper left hand side. The observation stations are marked by dots with identifying letters of their names alongside—thus W for Wick, RS for Royal Sovereign, V for Valentia . . . and so on. The reproduction of Fig. 44 may be, and is intended to be, photostated or otherwise suitably reproduced by the British yachtsman to a convenient enlarged size—e.g. as above recommended, about 13" × 10"—but, of course, it must not be deformed in reproduction; i.e. the enlargement must maintain the relative scale in all directions, as will be the case with an ordinary photographic process.

A Practical Example. The simplicity and practicability of the method will be appreciated from the practical example now to be given of the way in which an actual broadcast

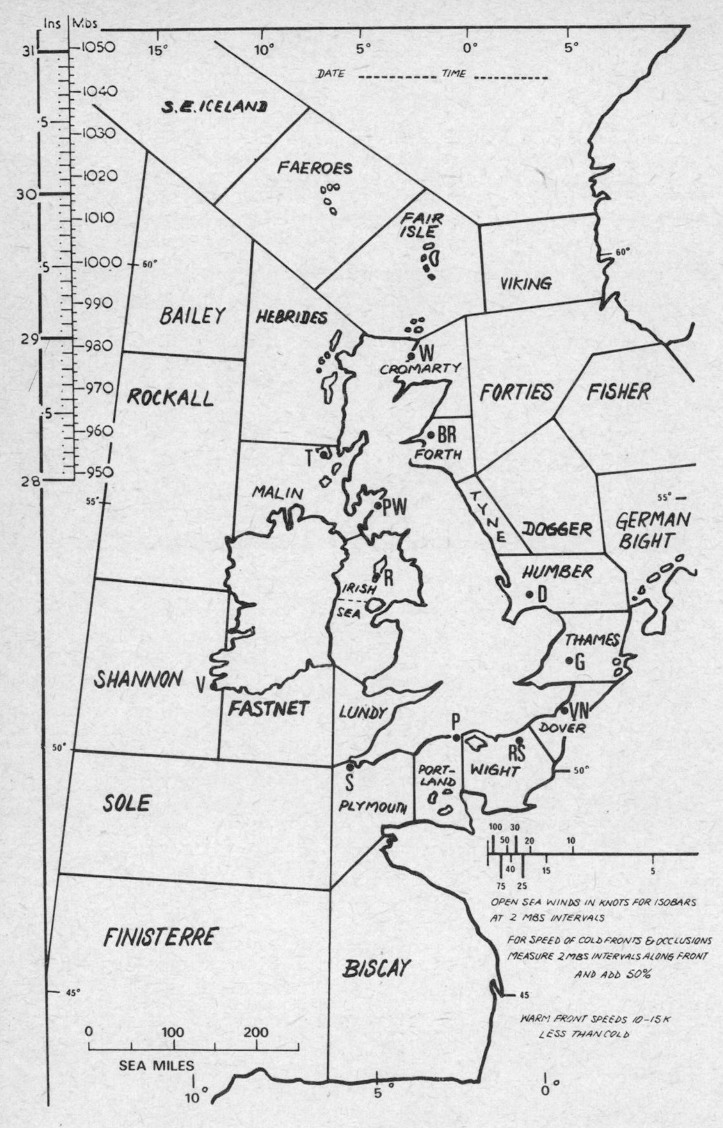
Ins
Mbs
31
1050
1040
1030
1020
30
1010
1000
990
29
980
970
960
28
950
15°
10°
5°
0°
5°
DATE ---------- TIME ----------
S.E. ICELAND
FAEROES
FAIR ISLE
VIKING
60°
BAILEY
HEBRIDES
W
CROMARTY
FORTIES
FISHER
ROCKALL
BR
FORTH
T
MALIN
55°
PW
TYNE
DOGGER
GERMAN BIGHT
R
IRISH SEA
HUMBER
D
THAMES
G
SHANNON
V
FASTNET
LUNDY
VN
DOVER
P
RS
WIGHT
50°
S
PORT-LAND
PLYMOUTH
SOLE
100 30
50 20 10
40 15 5
75 25
OPEN SEA WINDS IN KNOTS FOR ISOBARS AT 2 MBS INTERVALS
FOR SPEED OF COLD FRONTS & OCCLUSIONS MEASURE 2 MBS INTERVALS ALONG FRONT AND ADD 50%
FINISTERRE
BISCAY
45°
45
WARM FRONT SPEEDS 10-15 K LESS THAN COLD
0
100
200
SEA MILES
10°
5°
0°

FIG. 44.

can be used. The broadcast in question, transmitted at 0630, is as follows:—

Gale Warnings are in operation for all sea areas except Biscay and Finisterre.

The General Synopsis at Midnight. A vigorous depression of 957 millibars centred near Stornoway was moving South-East at about 30 knots. A warm front extended from the centre through the Firth of Forth to Spurn Head with a cold front lying along the West Scottish Coast and across N.W. Ireland. The centre of the depression is expected to be over South Denmark at midnight tonight.

Shipping Forecasts for the next 24 hours:

VIKING. Wind South-East, force 6 to gale 8 becoming East and later North. Occasional showers otherwise good visibility.

FORTIES, CROMARTY. Wind blowing anti-clockwise around depression force 6 to gale 8 soon becoming North and increasing to severe gale 9 and perhaps storm 10 at times. Occasional rain otherwise moderate visibility.

FORTH. As Humber.

TYNE. As Humber.

DOGGER, FISHER, GERMAN BIGHT. Wind South-West force 7 or gale 8 and severe gale 9 in Dogger and German Bight, becoming anti-clockwise around depression moving through areas and later North gale force 8 or severe gale 9 and perhaps storm 10 at times. Periods of rain otherwise moderate visibility.

HUMBER, FORTH, TYNE, SMITH'S KNOLL, THAMES, DOVER, WIGHT, PORTLAND, PLYMOUTH. Wind South-West to West gale force 8 or severe gale 9 becoming North-West to North. Rain at first, occasional showers later, otherwise moderate visibility.

BISCAY, FINISTERRE. Wind South-West to West force 5 to 7 but North-West force 4 in South Finisterre at first, becoming North-West force 6 or 7 over whole of area and perhaps gale 8 in places. Occasional rain otherwise good visibility.

SOLE, LUNDY, IRISH SEA, FASTNET, SHANNON. Wind West to North-West gale force 8 or severe gale 9 and storm 10 in places becoming North-West to North. Occasional showers otherwise moderate visibility.

ROCKALL, MALIN, HEBRIDES, FAIR ISLE, BAILEY, FAEROES. Wind North to North-East force 7 to severe gale 9 and storm force 10 in places at first. Occasional showers otherwise moderate visibility.

SOUTHEAST ICELAND. Wind North force 7 to gale 8. Occasional showers otherwise good visibility.

Weather Reports from Coastal Stations for 0500 *hours G.M.T.*

WICK. Wind North-East force 8. Continuous heavy rain. Visibility 2 miles. 958 millibars. Falling slowly.

BELL ROCK. West-South-West force 6. 22 miles. 957 mbs. Falling.

DOWSING. South-West force 8. Continuous moderate rain. 2 miles.

GALLOPER. South-West force 9. Continuous heavy rain. 2 miles.

VARNE. South-West force 9. Continuous rain. 4 miles. 979 mbs. Falling rapidly.

ROYAL SOVEREIGN. South-West force 9. Continuous heavy rain. 5 miles. 982 mbs. Falling rapidly.

PORTLAND BILL. West-South-West force 11. Continuous heavy rain. 4 miles. 979 mbs. Falling very rapidly.

VALENTIA. North-West by West force 7. Rain showers in past hour. 5 miles. 986 mbs. Falling.

RONALDSWAY. West-South-West force 8. 8 miles. 964 mbs. Falling very rapidly.

PRESTWICK. South-West force 4. Moderate rain shower. 3 miles. 958 mbs. Falling very rapidly.

TIREE. North by West force 7. Moderate rain shower. 3 miles. 960 mbs. Steady.

SCILLY. West-North-West force 8. Intermittent slight rain. 4 miles. 983 mbs. Falling very rapidly.

Plotting. Fig. 45 shows what may be termed "stage 1"—the plotting of the information as it comes in—and is finished by the time the broadcast ends. As there are gale warnings for all areas except Biscay and Finisterre, this fact is written on the chart as quicker and simpler than putting the letter G in every area except these two. The letter L with 57 in it records the depression centre, and the arrow, with 30 alongside, records the expected movement of the centre. The heavy black curved lines show the front as given by the forecaster. L′, with 2400 under it, records the forecast position of the centre at the next midnight. "Own ship" observations—the ship was very wisely tucked up in har-

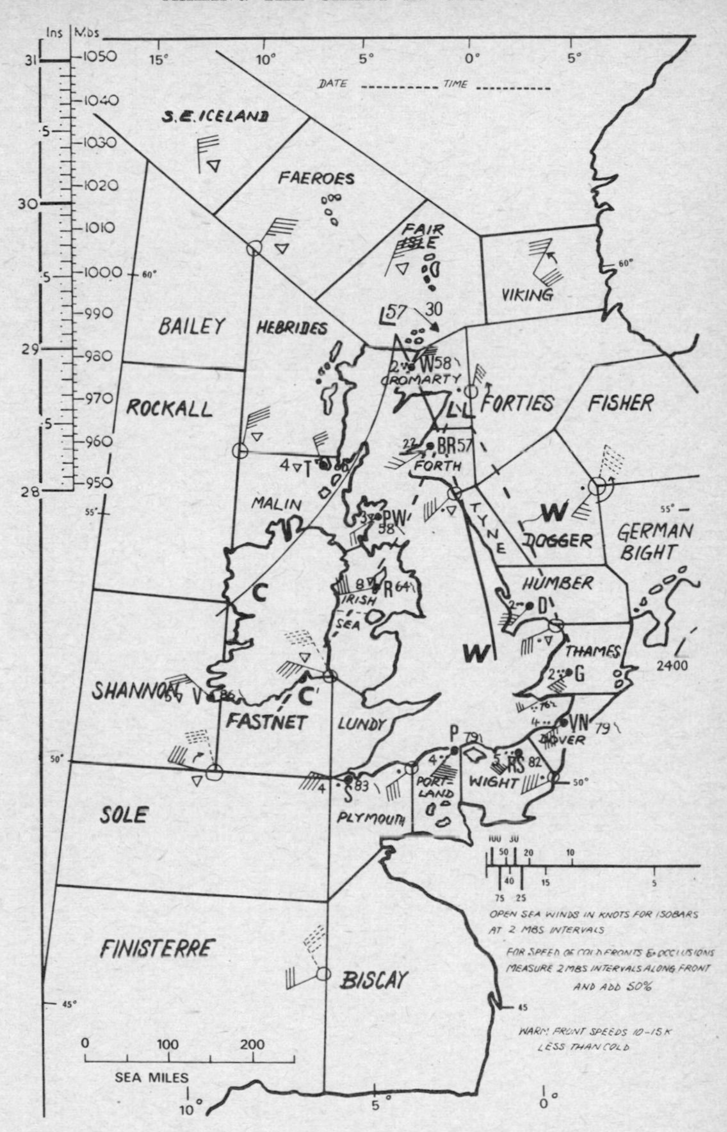

Ins
Mbs
DATE
TIME
S.E. ICELAND
FAEROES
FAIR ISLE
VIKING
BAILEY
HEBRIDES
ROCKALL
CROMARTY
FORTIES
FISHER
FORTH
MALIN
TYNE
DOGGER
GERMAN BIGHT
HUMBER
IRISH SEA
THAMES
SHANNON
FASTNET
LUNDY
DOVER
PORTLAND
WIGHT
PLYMOUTH
SOLE
FINISTERRE
BISCAY
OPEN SEA WINDS IN KNOTS FOR ISOBARS AT 2 MBS INTERVALS
FOR SPEED OF COLD FRONTS & OCCLUSIONS MEASURE 2 MBS INTERVALS ALONG FRONT AND ADD 50%
WARM FRONT SPEEDS 10-15 K LESS THAN COLD
SEA MILES

FIG. 45.

bour in weather like this—are shown near Chatham. The ship had South-West Force 6, continuous heavy rain, barometer 976 falling fast. There was the same forecast for a number of areas. When this occurs—and it is usual—it is often convenient to put one set of symbols on the common boundary where the areas meet, as this saves time in writing. Thus, for example, a single set of symbols is shown where the areas "Dogger", "Fisher" and "German Bight" meet and the same expedient is adopted elsewhere on the chart. Where the number of areas with the same forecast is large—in this case "Humber", "Forth", "Tyne", "Smith's Knoll", "Thames", "Dover", "Wight", "Portland" and "Plymouth" all have the same forecast—it is a good time-saving dodge to mark a few rings at convenient points distributed over the combined area and, at first, put in the symbols on one of them only. The symbols can be repeated at leisure on the other rings at the end of the broadcast.

Stage 2, which will not by any means always or even frequently be necessary, consists in moving the centres and fronts in accordance with the forecast so that they will "fit" the station observations. In the present example, with a deep depression moving at 30 knots, stage 2 is certainly necessary, for the "general synopsis" is for midnight and the station observations for 0500. The letter L and the two fronts which meet there are drawn for the midnight positions as given in the broadcast. When the broadcast is finished, the estimated position for 0500 is drawn in at LL and the fronts moved forward to correspond, as shown by the broken-line curves C′ and W′ in Fig. 45. As will be seen later, you may, when drawing the chart, have to move these positions a bit. It will be noted that the fronts are shown as simple unmarked lines, for it is perfectly obvious that the more Easterly one is warm and the more Westerly one cold.

Drawing the Isobars. Fig. 46 will serve to show how this is done. So as to make this figure as clear as possible for its intended purpose a good deal of the matter shown in Fig. 45 is omitted, though the estimated centre LL and the estimated frontal positions C′ and W′ are still shown. In practice, of course, when the plotting is done in china-glass pencil on the Perspex case of the chart, marks can be put anywhere—whether on the top of printed marks on the chart or not—without confusion, and Figs. 45 and 46, being in black and white, appear a good

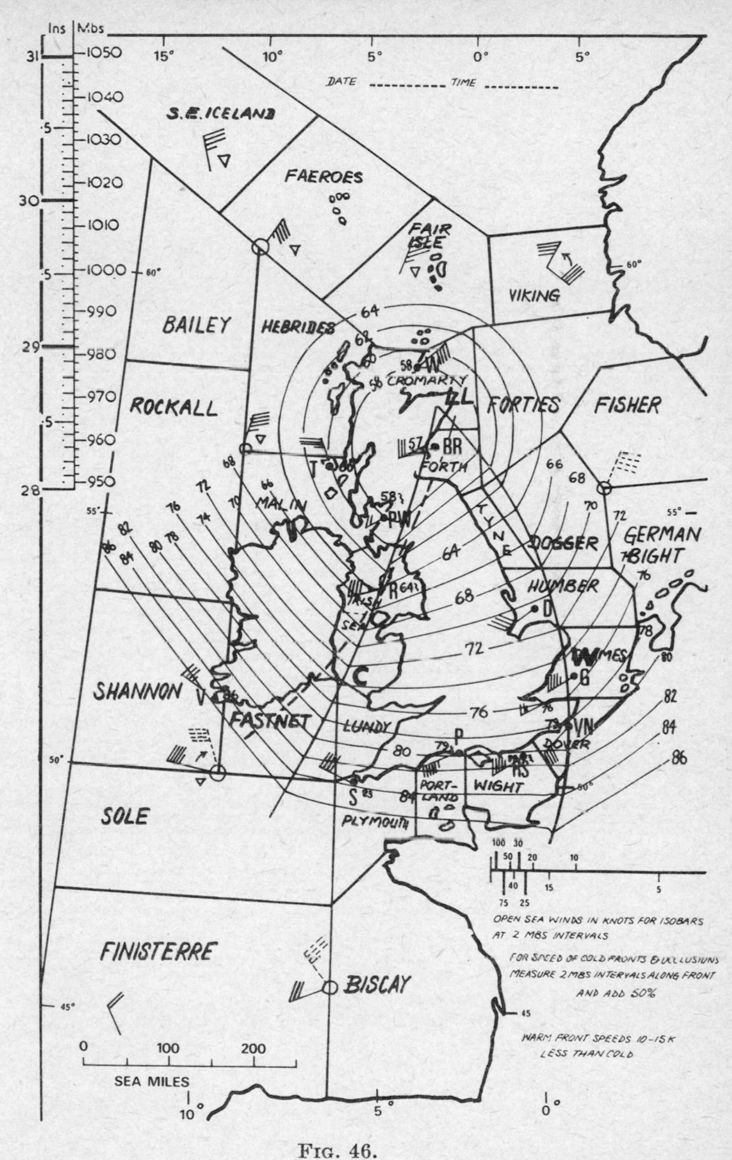
Ins
Mbs
DATE
TIME
S.E. ICELAND
FAEROES
FAIR ISLE
VIKING
BAILEY
HEBRIDES
CROMARTY
ROCKALL
FORTIES
FISHER
FORTH
MALIN
TYNE
DOGGER
GERMAN BIGHT
HUMBER
IRISH SEA
SHANNON
FASTNET
LUNDY
DOVER
PORTLAND
WIGHT
SOLE
PLYMOUTH
FINISTERRE
BISCAY
SEA MILES
OPEN SEA WINDS IN KNOTS FOR ISOBARS AT 2 MBS INTERVALS
FOR SPEED OF COLD FRONTS & OCCLUSIONS MEASURE 2 MBS INTERVALS ALONG FRONT AND ADD 50%
WARM FRONT SPEEDS 10-15 K LESS THAN COLD

FIG. 46.

deal more confused than would be the case in practice. Coming now to the drawing of the isobars, examination of the chart in Fig. 46 shows Bell Rock with 957 mbs., falling, and a S.W. wind; Wick, with 958 mbs., falling, and a N.E. wind; Tiree with a steady barometer at 960 mbs. and a N. × W. wind; and Prestwick with a falling barometer at 958 mbs. These four stations, with their wind directions, pretty well define the 958 mb. isobar, which can be drawn in and is shown as the innermost closed isobar over N. Scotland in Fig. 46. As will be seen, it satisfies the requirement, as any isobar in the Northern hemisphere must, of running more or less in the wind direction (within a compass point or so) with low pressure on the left and high pressure on the right.

It will be noted that the position of the 958 mb. isobar indicates that the centre of the depression is not moving quite as fast as was expected and that our letters LL are a bit too far to the S.E. The comparatively light wind (Force 4) at Prestwick is obviously due to local shelter, and the shift of wind between it and Ronaldsway (on the one hand) and Tiree (on the other) indicates that the position of the cold front in that part of the chart is about right, though it should originate in what now appears to have been the centre of the depression at 0500. The steady barometer at Tiree, behind the cold front (where one would expect a rising glass), is pretty obviously due to the fact that the depression is still deepening fast—as indicated by the rapidly falling barometer readings at Ronaldsway and Prestwick, both in the warm sector—and this deepening is balancing out the rise which would otherwise be occurring at Tiree.

The pressure of 976 mbs. at "own ship" (Chatham), 979 mbs. at Portland Bill and 983 mbs. at Scillies, all with South-West winds, gives a good lead to the run of the isobars around 980 mbs., but the pressure of 986 at Valentia lands us in some difficulty, with the chart as so far drawn, in making the 980 isobar pass on the low-pressure side of Valentia, as, of course, it must. We note that Valentia's wind has gone right round to the North of West and that it has showers, both of which facts are consistent with the cold front having well passed Valentia. It looks as if we may have drawn the trailing part of the cold front too far back. We accordingly advance the S.W. part of the front into St. George's Channel, as shown by the full line in Fig. 46, leaving it in its original position where it crosses the 958 mb.

isobar. This, incidentally, makes it originate very nicely in what the 958 mb. isobar leads us to believe is the centre of the depression. In practice we would also shift the warm front a bit, curving it to go back into the middle of the depression, but this is not shown since to show it would complicate Fig. 46 too much. If, with the cold front thus moved, we try the 986 or the 984 isobar again, we find they draw in very satisfactorily, fitting pressures, winds and weather. Accordingly we draw in the 986 mb. isobar (say) through Valentia and on the high-pressure side of the Scillies (i.e. South of them), changing direction at the cold front. In Fig. 46 the 986 mb. isobar is partly drawn and marked 86. Portland Bill (979 mbs.) gives us the run of the 980 mb. isobar; "own ship" (976 mbs.) shows us where the 976 mb. isobar goes; and Ronaldsway gives us the intermediate isobar of 964 mbs. All these isobars are partly drawn in and marked with their values in Fig. 46. The remaining isobars can then be drawn in to fill the spaces left and the isobaric spacing checked on the wind scale to see that the winds are reasonably in line with the forecast—as they are. The result is an eminently practical and useful chart. The usual practice is to draw isobars at 2 mb. intervals, but when they are as exceptionally close together as they are in this example, 4 mb. or even larger spacing is wise. In Fig. 46 intermediate isobars spaced at 2 mbs. are indicated only close behind the cold front so as not to complicate the drawing.

Fig. 46 is a somewhat exceptional chart chosen to show what can be done even when there is a great number of closely spaced isobars present. Normally, thank goodness, isobars are not often anything like as closely spaced as these are in what was one of the most violent and vigorous depressions experienced in British waters, even in winter, for a long time. A normal summer chart will have only a small fraction of the number of isobars that there are in Fig. 46; the station pressures will be enough to enable three or four of them to be drawn in without difficulty; and the remaining isobars can be "filled in" in a minute or two.

I recommend this method of chart drawing as thoroughly practical and most useful. Incidentally a little practice leads very quickly to a familiarity with weather charts and their development, together with a natural "weather sense", which can hardly be obtained in any other way. Each chart drawn makes the next one easier. It is, I think, wise to practise the method ashore, before the season starts, comparing your charts with those in some newspapers (the *Times* in the United Kingdom) or, better still, having the official weather reports (in the United Kingdom the "Daily Weather Reports" of the Meteorological Office) sent to you for a few months so that you can compare your charts with the official ones while you are practising.

CHAPTER XII

STATE OF SEA: LOCAL WINDS

So far we have considered only general weather charts and large area forecasts obtained therefrom, and have ignored local variations and effects. Obviously, however, local topography may make the weather in some particular small area quite different from what it is elsewhere. While a general area weather forecast has a great probability of being right for the open sea, the moment one is near land local effects must be remembered, and common sense used to estimate the way in which such a general area forecast is likely to be modified by local effects in some particular small part of that area. The most obvious of these local effects—so obvious indeed that it hardly needs pointing out—is shelter from wind and sea. There may be a forecast of a gale and very rough sea for some sea area, but although the forecast may be right you will not get either close up in the lee of a big headland.

State of Sea. The relations given in the table in Chapter II between wind strengths and accompanying states of sea apply only to deep water in the open sea and with nothing much in the way of tidal streams flowing. There are, however, four main factors which may cause the state of sea kicked up by a given wind strength to differ widely from that given in the table. These four factors are "duration" (i.e. length of time the wind blows), "fetch", depth of water, and stream.

Duration. Modern research into the generation of waves by wind has led to the conclusion that, for winds of about Force 3 or more, the sea in deep water will gradually build up to a limiting figure as given by the table in Chapter II if the wind blows long enough and over a sufficient area of sea. During this "building up" period the waves get higher and longer, and move faster until their speed is about the same as that of the wind itself. The stronger the wind the longer is the building-up period, and the higher and longer are the waves finally built up. Once, however, the limiting figure is reached the waves do not get any higher or longer however long the wind blows. From the practical point of view the little ship is mainly concerned with the height of the waves rather than their length, since any open-sea waves worth bothering about are very long compared to the length of a yacht. In round figures it may be stated, with sufficient accuracy for practical purposes, that for winds of about Force 5, or more, the sea will reach about one-half its final maximum height in 6 hours, about three-quarters of its final height in 24 hours, and its final height in 48 hours.

Fetch. *Fetch* is the length of sea path over which a wind blows. Everybody knows that a really powerful wind off the land will produce no sea at all close in-shore, but it is not generally realised how long a fetch is necessary before the sea can be built up to even a large fraction of its maximum final height. Again, giving round figures, winds of Force 5 and above build up to about one-half the maximum height with a fetch of 50 miles and to about two-thirds the maximum with a fetch of 100 miles. Opinions differ as to how long a fetch is necessary to achieve the maximum, but it is almost certainly well in excess of 300 miles, and probably nearer 500 miles.

Depth of Water. Contrary to popular belief, the height of a wave does not change much when it comes into shallow water. What happens is that when a wave comes into a depth which is about one-half the wave length, friction of the bottom slows it up and also brings down the wave length. Accordingly the number of waves in a given distance increases and, since the height remains much the same, they become steeper and therefore, so far as the little ship is concerned, more vicious. The sea produced by a Force 5 wind blowing for about 6 hours over a fetch of, say, 50 miles will show about four crests to the cable, so that, in these circumstances, the wave length is about 150 ft. and a depth equal to the half wave length is 75 ft.—say 12 fathoms. If, with such a sea running, you enter depths of less than about 12 fathoms, you can confidently expect the sea to get increasingly steep and vicious as the depth decreases. What I think of as the "vice rule"—depth below a half wave length—is easy enough to remember and to practise in a rough-and-ready sort of way, for it is not difficult to assess wave lengths by eye, knowing the length of your own ship.

The other important effect of shallow water is to increase the speed with which a sea kicks up—and, for the matter of that, subsides. This is one of the factors which give places like the Isselmeer in Holland (the Zuyder Zee as it used to be called), and estuaries when the wind blows straight into them, their unenviable reputation as far as little ships are concerned. The figures above given for "duration" and "fetch" do *not* apply to shallow water.

It is also as well to remember, when it comes to such things as entering a harbour with a bar outside, that a wave will break if the depth under it is not more than about 1½ times its height—i.e. a 6 ft. wave will break in 9 ft. of water.

Stream. The effect of stream (tidal or other stream) upon sea is obvious enough. It is *relative* movement of wind to sea surface which produces waves. Clearly, therefore, if the sea surface is moving in the same direction as a given wind the relative wind is less than if it is moving in the opposite direction and the sea kicked up will be less. The *relative* wind produced by a 15-knot wind in the same direction as a 5-knot tidal stream is only 10 knots, which produces very little sea even in shallow water. The same wind meeting a 5-knot stream in the opposite direction produces a relative wind of 20 knots with correspondingly more sea. Since fast tidal streams are phenomena associated with shallow water and constricted fairways, it is, in general, only in such places that tidal streams need be considered, meteorologically, by the weather-wise mariner.

Swell. Waves caused by a wind blowing somewhere else and which have travelled out of the wind area are known as *swell.* Meteorologically, swell is of very little importance, for, despite widespread belief to the contrary, modern research has shown that it is practically only in the case of a low-latitude tropical storm that the swell travels ahead of the wind. Practically speaking, in temperate latitudes, if the wind is coming your way you will get it, and the sea which accompanies it, without any warning by swell, and therefore swell without much wind, though uncomfortable, generally only means that somebody somewhere else is having a bad time.

Land and Sea Breezes. The intelligent anticipation and use of *land and sea breezes* can be very helpful to the mariner coasting under sail, often enabling him to get a better wind by closing the land, or leaving it, at the proper times. They are by-products of the fact that land can heat up and cool off quite rapidly whereas sea temperatures do not alter quickly. Consider what happens when the land along a coast gets hot during a hot summer's day and suppose, for the moment, that there is no "meteorological" wind—i.e. that, from the weather chart alone, one would expect a flat calm. Fig. 47a illustrates this condition. The air over the heated land is warmed and rises—usually about 500 to 1000 ft. —and its place is taken by air which blows in off the cold sea so that there is a local circulation of air as shown by the arrows, giving an onshore or "sea breeze". At night, if the land gets colder than the sea, there is a similar circulation, but in the opposite direction, as shown in Fig. 47b, with an off-shore or "land breeze". Both effects are purely local and the sea breeze is seldom felt more than 10 miles or so off the coast: the land breeze seldom extends more than 5 miles from the coast. In

FIG. 47a. FIG. 47b.

British waters these breezes seldom exceed Force 3, though in the Mediterranean and the tropical waters the sea breeze sometimes reaches as much as Force 5. The sea breeze, which is generally much the stronger of the two, usually starts about 5 hours after sunrise, reaches its maximum by mid-afternoon and dies away towards sunset. The land breeze usually sets in a couple of hours or so after sunset and dies away around sunrise. Since both breezes are by-products of land temperature change, they will not occur if there is much cloud cover, for cloud acts as a "blanket" to prevent quick heating or cooling of the land. Clear skies, or fairly clear skies, are therefore necessary for them to occur.

Although the forces causing land and sea breezes tend to make them blow pretty well at right angles to the coastline, it will be apparent that if there is a "meteorological" wind force as well the overall result will be compounded of the two wind forces. Thus, in the case of a coastline running North and South and an Easterly sea breeze of 10 knots, if the pressure gradient as shown by the chart gives a Westerly surface wind of 10 knots, the overall result will be a flat calm. If, with the same sea breeze, the "met." wind is Easterly, 10 knots, the overall result is a 20-knot East wind near the coast. If the "met." wind and the land or sea breeze are at an angle to one another, the resultant wind will be between the two.

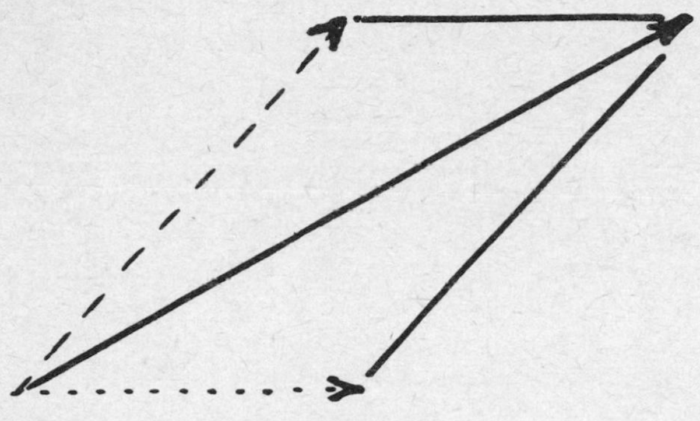

FIG. 48.

This is illustrated by Fig. 48, in which the dotted-line arrow represents, by its direction and length, a sea breeze and the dashed-line arrow similarly represents a "met." wind. The full-line arrow represents the resultant actual wind.

Katabatic and Anabatic Winds. Another form of local wind regularly experienced in some localities is the so-called *katabatic wind.* If high ground which slopes down to the coast cools off on a clear night in summer, or if it is covered with snow in winter, it will cool the air in contact with it, and the cooled and therefore heavier air flows down the slope by gravity and blows a short way out to sea. The effect is very local and the wind thus caused does not extend, as a rule, more than a short distance from the coast, nor does it, in home waters, attain much force—hardly ever above Force 3. Such an off-shore wind is called a katabatic wind and is commonly experienced near mountainous coastlines. Fig. 49 illustrates a katabatic wind by arrow-headed broken lines, and, as in the case of land and sea breezes, it is likely to be compounded with a "met." wind.

Anabatic winds are the opposite of katabatic, occurring when, on a hot day, air on a sloping land surface is heated and thus

FIG. 49.

caused to run up the slope. From the point of view of the seaman, however, they are of little, if any, importance and seldom, if ever, give rise to appreciable wind over a sea surface even when the slope runs right down to the sea.

A FINAL WORD

Meteorology is still far from being an exact science, for the interplay of forces which occur in the mechanism of weather is so intricate, and the theatre in which that interplay takes place is so vast, that even with the elaborate organisation of international observation which has been built up over the years it still remains true that any weather chart *may* be wrong. Indeed, it is probable that something is missed in *every* chart. Nevertheless, the modern meteorologist may justly claim to have advanced a very considerable way along the long road to exact knowledge, and every day that passes sees him a little farther along that road. The average accuracy of his forecasts nowadays is pretty high, but if those forecasts are to be used to good practical effect in the little ship at sea the master *must* understand at least the main underlying principles of the subject so as to be able to interpret the forecasts intelligently. I am sure that the weather man is blamed far more often for the errors of the listener than for his own errors. I am sure, also, that the practice of meteorology at sea, to the utmost extent that the limited facilities of the little ship will permit, leads, as nothing else can do, to the cultivation of weather wisdom and of a "weather eye" that are so essential to the good seaman, especially the seaman in the little ship. So make your charts and watch the sky and the barometer, and, despite your mistakes—even your spectacular ones—let no wiseacre deter you by silly jeering about "seaweeds". Few subjects are more interesting, as well as useful, than meteorology at sea, and, incidentally, few practices offer a greater reward in sheer beauty than observation of the ever-moving, ever-changing pictures and shapes in the sky.

So good luck with your forecasts. They'll get better as you go on.

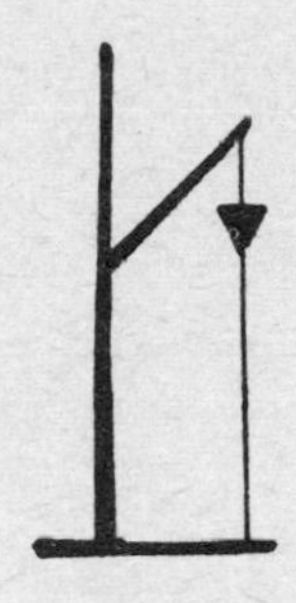

ADDENDUM: SOME WEATHER SIGNS

I am told that tradition requires any book on meteorology to contain something about the more popular "weather signs", and accordingly I bow to tradition to the extent of providing a table of some of them, with comment, in this addendum. Frankly, I do not attach much importance to any of them, because even the good ones only amount to observation of one result which may have many causes. However, for what they are worth, here they are:

Sign	*What of*	*Comment*
Mackerel sky	Bad weather	Hardly ever true in British waters. Not better than an even-money bet around latitudes 35°–45° N. Tossing a coin is as good as an even-money bet!
Red sky at sunset	Good weather	Fair. It depends on the shade. A gentle red or pink at sunset or sunrise indicates dry air and to that extent accompanies good weather, but fiery reds—or indeed any violent colouring—indicates moist air in the distance and is therefore often a close fore-runner of bad weather.
Red sky at sunrise	Bad weather	
Halo round the moon	Bad weather	Good. A halo is caused by cirro-stratus and therefore its presence often indicates the advancing front of a depression.
Mare's-tails	Bad weather	Good. Cirrus is pre-frontal cloud, but there are other causes of cirrus.
Seagulls come inland	Bad weather	Seagulls, being sensible birds, certainly fly in when it is blowing hard at sea. There is no evidence, however, that they do it *before* the wind comes. All the other numerous "animal behaviour" signs can be similarly classed as old wives' tales.
"Dog" to lee	Good weather	Not bad. A "dog" is a rainbow and is seen in showery weather. These "signs" assume the weather to move Eastwards (as it generally does) and the wind to be in the Western half-circle (as, in British waters, it mostly is). A "dog to lee" therefore means one to Eastwards, and since the rainbow is usually in the cold air behind a cold front or occlusion this means the front has passed. There are, however, much better means of knowing this!
"Dog" to windward	Bad weather	
Rain from the East A full day at least	Self-evident	Pretty good. An east-moving front is likely to be slow-moving and, indeed, may stop and go back. Rain from it is likely, therefore, to last a long time—occasionally for days.
Rain before seven, Fine by eleven	Self-evident	Pretty good. This old saw is a statistical gamble on the fact that warm-front rain often lasts about 4 hours or so.
Exceptional visibility and a "hard" horizon	Bad weather	Fairly good. The cause is rather obscure, but it appears that moist air in the upper levels ahead of a warm front often produces this effect, probably by some sort of light refraction.

Sign	*What of*	*Comment*
Change of moon, Change of weather	Self-evident	This belief is so widespread that a careful statistical study extending over the reports of many years was carried out to check it. The study produced no evidence in support.
When the sea-horse jumps Look to your pumps	Self-evident	The sea-horse is the porpoise. He often jumps, being a playful beast, but he is no forecaster. However, since it is always wise to look to your pumps the advice is sound, though useless for weather forecasting purposes.
If November ice will bear a duck Rest of winter's nowt but muck	Self-evident for Northern latitudes where winter starts in November	This, like other purported "long-range forecast" indications, is quite useless. The sad fact is that nobody yet knows enough to forecast a long time ahead in variable-weather latitudes.

INDEX

INDEX